HOW WE HEAR MUSIC

THE RELATIONSHIP BETWEEN MUSIC AND THE HEARING MECHANISM

HOW WE HEAR MUSIC

THE RELATIONSHIP BETWEEN MUSIC AND THE HEARING MECHANISM

James Beament

THE BOYDELL PRESS

First published 2001
The Boydell Press, Woodbridge
Reprinted in paperback 2003

ISBN 0 85115 813 7 hardback
ISBN 0 85115 940 0 paperback

The Boydell Press is an imprint of Boydell & Brewer Ltd
PO Box 9, Woodbridge, Suffolk IP12 3DF, UK
and of Boydell & Brewer Inc.
PO Box 41026, Rochester, NY 14604–4126, USA
website: http://www.boydell.co.uk

A catalogue record for this book is available
from the British Library

Library of Congress Cataloging-in-Publication Data

Beament, James, Sir.
  How we hear music: the relationship between music and the hearing mechanism /
James Beament.
    p. cm.
  Includes bibliographical references and index.
  ISBN 0–85115–813–7 (alk. paper)
  1. Musical perception. 2. Hearing. 3. Music – Acoustics and physics. I. Title.
  ML3838 B43 2001
  781'.11–dc21   00–067512

This publication is printed on acid-free paper

Typeset by Joshua Associates Ltd, Oxford
Printed in Great Britain by
St Edmundsbury Press Ltd, Bury St Edmunds, Suffolk

Contents

Figures

Tables

Preface

Music depends entirely on the sense of hearing, and this book is literally about how we hear it. During the past fifty years there have been spectacular advances in our knowledge of how the ear and hearing system work. In its advanced form it is a large, extremely complicated subject and really a closed book to all but specialists in that field of study. One can, however, extract a simplified explanation of the mechanism, to which basic musical phenomena can be applied. But add to that our modern understanding of evolution and behaviour, of how advanced animals including ourselves use their hearing, together with acoustics, and the mass of fact and belief about hearing music, and one is faced with a huge body of uncoordinated and sometimes conflicting material.

In such a situation in science, and it appears equally true of music, it is often fruitful to go back to first principles. So the book begins by discussing the origin and early evolution of simple 'western' tonal music, which appears to be almost universally accepted and acceptable. No one knows how music originated. I suggest that it started with experiments with artefacts – with instruments, and not with the human voice. This is not unfounded belief, for the later chapters appear to substantiate the assumption, and if it runs counter to your current belief, I ask you to give it the benefit of the doubt until you have read all the arguments.

A consideration of the evolution of simple music from first principles produces a list of basic questions about intervals and scales, tone, dynamic, harmony, time and so on. And as the discussion develops it leads amongst other things to the conclusion that the harmonics of musical sounds, which are the basis of so much theory about music, did not and cannot play the role which has been so widely attributed to them ever since they were revealed by Helmholtz in 1870. I then examine whether the hearing mechanism provides some form of answers to the questions and conclusions. I believe that it does and that it produces a different view of the basis of some fundamental features of music to those which are commonly held. It also provides a cogent explanation of why our hearing mechanism behaves as it does, and therefore why we receive the sensations of music in the form we do.

I have arranged the book in this way because from long experience in teaching and examining Cambridge music students in acoustics, I know that there is little point in trying to interest most musicians in acoustics for its own sake. Similarly, there may be enquiring minds who are interested in the approachable account of our hearing mechanism I have given, but the same generality applies. The only thing one may reasonably expect to interest most musicians is whether either offers an explanation of the musical phenomena with which they are concerned. In other words, I start with music and relate it to science, rather than the other way round. I also know that mathematical equations are of no help to most people; many scientific things can be explained without them, and this book does so.

It may reassure those who have had difficulty in understanding writings about music, whether in popular articles or serious works, that I do define precisely the way in which I use common terms such as beat, note, pitch, timbre and tone. These restricted definitions may not correspond to your ideas of them, but you will always know what *I* mean by them.

I hope it will also reassure readers that I am a practising musician who plays jazz and serious music and composes both kinds, as well as a scientist.

Finally, I believe that the mainspring of music has been the production of pleasurable sound sensations. Even that is not now universally accepted, and if you are one of those who think otherwise, then I'm afraid this book may not be for you. But I hope it will be of particular interest to practising musicians and teachers, as well as to acousticians, those studying the psychology of music, or those involved in electronic music and recording.

Acknowledgements

I am of course indebted to the scientists who have investigated hearing. I have learned much over the years in conversations with many amateur and professional musicians. I have discussed various problems with Christopher Beament, Clifford Bibby, Brian Callingham, Christopher Longuet-Higgins, Giulia Nuti, Robin Walker, Richard Wilson and with Malcolm Macleod and Ben Milstein, who also read the manuscript. My wife, Juliet Barker, has contributed much from her own extensive experience of playing and making instruments, and tempered some of my more outspoken remarks. I benefited in very different ways from the two referees who read the book. And I am most deeply indebted to my colleague Dennis Unwin, who unstintingly read several versions of the manuscript, constructively criticised every draft, and made valuable contributions to the material, as well as carrying out experiments at my suggestion and making equipment for some of my experiments. I also thank Richard Barber, Caroline Palmer and the other staff of Boydell & Brewer for their continuous help throughout the preparation of this book.

Cambridge   J. B.
June 2000

1. Preliminaries

1.1 Musical arithmetic

When I was a small boy in the 1920s, any conversation about music interested me, and I took every opportunity to ask questions. 'What do you mean by a violin is tuned in fifths?' My friend's mother adjusted the pegs and played the open strings. 'Those are fifths,' she said. 'Why are they called fifths?' 'I don't know. That's what they are always called.' My cousin had had piano lessons. He played middle C on the piano and counted along the white keys 'One, two, three, four, five,' to the G. 'That's why it's a fifth.' He continued from the G to the C above, counting again. 'And that's a fourth.' 'So you just count along the white ones?' 'Yes,' he said, and played a series of fifths: CG, DA, EB, FC, GD, AE, paused, and played BF♯. 'But why did you play the black one?' 'I'm not sure,' he said, and then with a flash of inspiration, 'If you count all the keys, black and white [he demonstrated], a fifth is always eight of them.' 'And a fourth?' He counted again. 'That's always six of them.' Then he added 'And a fifth plus a fourth makes an octave – that means eight, you know.' I was good at arithmetic. 'So there's fourteen notes altogether in an octave.' 'I don't think so, [counting] only thirteen. Anyhow, an octave's twelve semitones.' And if the White Rabbit had hopped out of the piano at that moment it would not have surprised me.

None of it made any sense. Afterwards I went to the piano on my own and counted along five white notes for a fifth. I was happy with CG, DA and so on but this time I counted and played BF instead of BF♯. It didn't sound right. BF♯ did. So a fifth was something I heard from the right pairs of piano notes. It was the something I'd heard from the pairs of violin strings, though it was a completely different kind of sound to the piano; and a violinist could adjust the strings until she heard that they were producing this something. My hearing knew that some sounds were fifths and some were not, and so did the violinist's. I don't think my cousin's hearing knew. Musical hearing can recognise many different features of sounds. Before we can discuss how our hearing system might achieve it, we need to know what features of the sounds our hearing is recognising. This is an odd situation because in a way, hearing knows and we don't. And thank Heaven it does.

1.2 Musical sensations

What everyone experiences when sound is received by the ears are sensations. And music is a series of sensations; it started as sensations and however complex and sophisticated it has become, it is still just a set of sensations. Musicians have devised a very large number of overlapping terms which are simply names for things they can distinguish, or believe they can, in sensations. Most of the terms tell us little if anything about what it is in the sounds which are producing those features of the sensations. Most musicians: composers and players don't know. Composers have usually worked entirely in terms of the sensations which they liked and hoped other people would like. Most listeners have no idea, though they usually know what they like and dislike. Describing sensations isn't easy, sound sensations happen to be difficult, and some features of music sensations are virtually impossible to describe, but with considerable experience, one comes to appreciate that many of the terms do relate to real phenomena; a few of them are pure delusion.

So we have first to discover what characteristics of sounds are responsible for the features of the sensations to which musicians can give names. Then we may be able to see if they can be connected with how our hearing machinery operates. What is 'fifth'? What are we identifying without knowing what it is? How did we select it in the first place? Is our hearing machinery playing a role in this? When musical sounds were analysed over a hundred years ago, the components which were discovered called harmonics, appeared to offer a simple explanation for many things about music. Since then, physicists have analysed and described the sounds in great detail. But the sounds in the air are not a description of the sensations. There are what appear to be very obvious things in the sounds which do not appear to be in the sensations and very small things which do. One cannot assume that an analysis of a musical sound shows what we can hear; it has also led people to believe they can hear things which they can't. Suppose, for example, that we could not hear harmonics; some of the most basic explanations of music would collapse. How accurately can we hear the sounds of a fifth? Fortunately, less exactly than many musicians believe they can, because if they could, music would have to be played so precisely, it would not be practicable. But those things depend on our hearing machinery, and not on descriptions of the sounds.

For our purposes, what we need to know about the sounds can usually, I believe, be described in relatively simple ways. The features of the sounds which create the sensations of some of the *intervals*, the sensations which can be identified as common to pairs of notes regardless of the instrument on which they are produced, are not difficult to describe, and they are amongst the most important sensations in all music.

It might not help many musicians if the sensations of intervals had been given names which reflected the features of the sounds which do produce those sensations. But it might be better to have meaningless names than ones which produce the mayhem recounted in the first paragraph. The traditional terms confuse the logic which our hearing has selected in the sounds, and can send people on the wrong track (see Appendix 1). There are reasons why a keyboard was laid out in its conventional form some centuries ago. If one counts along some of the white keys, the numbers correspond to the names given to the intervals, but it is that way round. The sensations of the intervals were recognised long before. These visual keyboard images are of no help to other players like violinists. An octave is an interval determined by and recognised by hearing two appropriate sounds; the suggestion that it is a fifth plus a fourth is a nonsense.

It is when we come to terms like 'tone' that we meet real difficulties. One may be sure musicians know what features of a sensation they are trying to describe as tone, but the variety of adjectives applied to the word indicates the problem, even if we take into account personal taste and belief; possibly the nearest phrase in ordinary language is the quality of the sound, which does not get us any further. Tone is an extremely important sensation. Orthodox musical instruments were entirely developed by musical makers using the sensation of tone as the criterion. They had no idea of what they were actually selecting. We do now know, but unfortunately acousticians jumped to the conclusion that therefore tone was a property of instruments. Some acousticians and many makers still believe that. The changes in instruments made it possible for *players* to produce sounds with different tone characteristics. We must discover what features of sound our hearing uses when the tone sensation is produced, and which ones do not play any role. The answer is rather surprising.

In broad terms, we can distinguish what kind of instrument is being played, the loudness, what we describe as tone and timbre, the sensations of intervals, and where sounds occur in a time system, all from the same sounds, as well as the direction from which the sound appears to originate. We cannot get any further without a rough idea of what sound is. We only need a rough idea, because it is not only pointless but also misleading to think that music can be described in precise physical terms. That originates in some persuasive Victorian explanations of musical sound and of music which have become embedded in the literature and used in teaching. The study of hearing upon which we are now embarking suggests that some of those ideas are no longer tenable.

If you are familiar with acoustics I suggest you read Sections 1.3, 1.4 and 1.5 rapidly, but please don't omit them, because they contain explanations of exactly how I use certain terms. I hope there will be some readers who are not so familiar with the material.

1.3 Sound

Sound is *vibration*; it is an unusual kind of vibration, and it is *not* like the picture often found in popular books, of ripples spreading out from a stone thrown in a pond. We shall be mainly concerned with sound travelling through the air. Air is a gas: many billions of billions of minute particles, called molecules, in every litre, which are evenly distributed in still air. Take a bicycle pump, close the hole by holding a piece of rubber or leather over it (don't just use your finger or you may burn it), push the plunger in a few centimetres and let go. The plunger comes back. You have compressed the air, pushed the molecules closer together, and when you let go they expand back and push the plunger out. We can apply that idea to a sound producer, for example, a drum. We hit the stretched skin and it vibrates; we can see and feel that the skin is vibrating in and out. Sound is always produced by something vibrating. As the skin moves out it pushes on the layer of air molecules immediately in contact with it. Momentarily it compresses that air, but unlike the air in the pump, that air can expand again straight away because it is not confined; in doing so it compresses the layer of air next to it. And that process of compressing the next and the next layer after layer of air goes on as a continuous event through the air, travelling away from the drumskin.

The drumskin immediately moves in the opposite direction: the second half of its vibration. This time it sucks on the layer of air and expands it; the molecules are separated more than they are in still air. And that lowering of pressure sucks on the next layer of air and so on. And since the drumskin goes on vibrating, the sound it produces consists of two things happening to the air. At any one place as the sound moves through the air, the molecules are moving backwards and forwards, and they are moving closer together and then further apart than in undisturbed air, and, therefore, the air pressure at that place repeatedly increases and decreases. The air as a whole does not move; it is only being vibrated in that way as the sound passes through it. I still find it remarkable that this process transmits virtually every minute detail of the complicated vibrations of the sounds of instruments and voices. And it is remarkably efficient. Very little energy is contained in the sound and very little of it indeed is used up in passing through the air. Far more is used up in making the drumskin vibrate. Of course this cuts both ways, because if the sound contains very little energy, it has very little to operate our hearing machinery, which has to be correspondingly sensitive.

Obviously, the bigger the vibration of the thing which pushes and sucks on the air, the bigger will be the movement back and forth and the size of the pressure changes in the air, and therefore the louder we hear the sound. But the connection between the size of the changes, and the sensation of loudness is not simple. That is discussed in Chapter 8. Physical devices can provide a crude measurement of noise levels, but they cannot measure the loudness of sounds, musical or otherwise, because they are sensations. As for the image of ripples on a pond, we now see why such an analogy is misleading. The water goes up and down as the wave moves; air goes backwards and forwards, in the direction in which the sound moves. Musical sounds are somewhat regular, but not as regular as those ripples; in other noises the movements are likely to be as irregular as waves on a rough sea. That suggests there is something special about musical sounds which we should be able to describe. In contrast, there is a remarkable dearth of words which describe anything common to the sensations of the vast variety of other noises we hear. Compare that situation with the characteristics which are shared by anything we can see. We can describe natural objects as round, straight, square, or blue, green, red, bright or pale, and so on. Round applies to many things, and so does green. We can't describe any sound sensation as having shape or colour (though some people try); our categories for natural sounds and many other noises, are few and very generalised; they are just low or high, loud or soft, a buzz, rattle or whistle. It is a human trait that if we can observe properties common to things, we do so, and we have with musical sounds. But we are virtually limited to describing every natural noise by what we think may have made the noise, and if we try to put them into categories, we class the things making the noises instead of the noises. Birdsong covers a huge variety of noises; to go any further we are reduced to naming the kind of bird, if we can. So why can we describe things which are common to different musical sounds in the way we do?

1.4 Pitch and frequency

Pitch is that property of a sound sensation which enables us to say whether it is high or low. We can describe a scream as high pitched or a dog's growl as low pitched. Broadly speaking, the faster the rate of vibration of the sound, the higher the sensation of pitch. But the only sounds about which we can judge the pitch of one to be only very slightly higher than another, or the same as another, are the ones we hear in music, because they produce the sensation of *steady pitch*, which is only produced when the vibration of the sound is sufficiently steady. All steady-pitch sounds except a synthetic one called a *pure tone*, consist of several rates of vibration. They would have those rates of vibration if we didn't exist. But they would not have pitch, which is a single sensation of highness or lowness. It is perhaps the biggest puzzle about the mechanism of our hearing system that we do obtain the single pitch sensation from a sound consisting of many different vibrations. I make a tentative suggestion about how that happens at the end of Chapter 9. But if it didn't happen, we would not have music.

The rate of vibration of a pure-tone sound is called its *frequency*, stated as the number of vibrations per second. If it consists of many rates of vibration it has many frequencies. Unfortunately, physicists also call rates of vibration the 'pitch'. In discussing hearing, we must restrict pitch to the meaning in the previous paragraph, the sensation, or there will be confusion.

When we tune instruments in an orchestra, the oboe player obtains a pitch from a tuning fork by hearing, adjusts his oboe so that he obtains the same pitch, plays it, everyone obtains that pitch and adjusts their instruments until they obtain the same pitch (or in a few cases a pitch derived from it one or two octaves below). Very few people know what the rates of vibration are. We assume the tuning fork has produced 440 vibrations a second, because that is stamped on the fork, and it produces a constant pitch sensation. There is an interesting feature of how we use our hearing in the way we tune instruments. Most people compare two pitches by listening to one and then obtaining the other. We remember one pitch and compare the other pitch with our memory of the first one. There are good reasons for doing it that way, rather than listening to them together.

The term now used by physicists for 'vibrations per second' is hertz (abbreviated Hz), as, for example, 440 vibrations per second is written as 440Hz. If pure tones are vibrating very rapidly, they can be given in kilohertz (kHz) meaning thousands of hertz; 2000Hz is 2kHz. A possible reason is that Hz means the same in all languages. We must not say that the pitch of oboe A is 440Hz. We will decide in Chapter 3 how we can refer to values of pitch.

1.5 Transients

We cannot compare the pitches of other noise sensations in the way we do musical sounds because all their vibrations are continuously changing, both in their rates of vibration and in the loudness of each of them, in a very complex way. Every noise changes in a sufficiently different way from any other, for us to be able to tell each of them apart. Although any kind of sound can be called a noise, and the word has several different connotations, a better word for all noises except a steady-pitched sound is *transients*. A transient means a changing sound, and it also includes the sense of something which appears and disappears rapidly, which is a useful reminder that all sound goes past our ears at a speed of about 330 metres (1100 feet) a second, and it is only available to our ears at the instant it passes us and then disappears. If we are to remember transients we have to remember the way the sensation changes instant by instant, and then recognise that same way when we obtain a sensation that changes in the same pattern. Many of them last for much less than a second. Most of us remember and can identify many thousands of transients: noises and words.

The entire use of our hearing as a means of obtaining information from sounds depends upon remembering, at least in the short term, some length of the sensations we receive. We can only recognise a word by remembering the sensation over the time it takes to receive it. We extend that to remembering a succession of words. It becomes an automatic feature of using our hearing. By the same process, we recognise that a pitch is steady. And our automatic memory provides a continuous memory of a succession of musically pitched sounds. The process is essential if music is to communicate anything to us. This leads to two conclusions. One of the two basic features of western music is that we compare one pitch with another (the other being time patterns). We must have sounds which are vibrating steadily in order to achieve that. Sounds with reasonably steady pitch can occur inadvertently in the natural world; the most obvious are the hums and whines of insect wing-beats. But it would be surprising if any noise made by a terrestrial animal for communication had sustained steady pitch, for the reason we have just seen. If an animal makes a noise purposely, it wants the noise to be recognised, and noises are recognised by their transient pattern, because that is the only thing which makes noises unique. Steady pitched sounds produced so that one pitch may be compared with another is a special human artefact (Beament, 1977). We can forget the fairytales about music starting with birdsong, brooks, wind in the trees and any other transient noises. The reason it has proved so difficult to synthesise speech or any natural noise effectively is that they are all extremely complex transients, and our hearing which judges them is also extremely sensitive to transients.

As we see in the next chapter, discovering how to make sound with an artefact was an accident. It may be doubted whether the fact that it made one steady-pitched sound was recognised as unusual. A single pitch does not make music. What the artefact made possible was selecting the sensation of pitches which make intervals. Music has been developed ever since using the sensations of steady-pitched sounds in intervals. It has intrigued inquisitive people from the earliest developed civilisations. But fifths and other interval sensations were discovered by hearing long before that, by people who had no idea that sound was a vibration, or that musical pitch was produced by a steady vibration. Pitched sounds are an artefact; a fifth is an invention of our hearing.

But there is also a paradox. If sounds are not changing, and lack the characteristic which makes transient sounds recognisable, can we recognise what is making a steady-pitched sound? Most readers will respond that of course they can recognise what instrument is being played. We return to that important question later. Because if we can't, that would make the pitch a far more significant feature of the steady sensation.

1.6 Auralising

I want to avoid the confusion which can arise because *notation* which means written instructions, is usually called 'music'. A few years ago, a member of the violin-making classes for amateurs which my wife runs, reached that pregnant moment when he completed his first violin, made more emotional because it was the 14th of July and he was French. He was a competent violinist and wanted the first thing it sounded to be his National Anthem. He could sing the Marseillaise unaccompanied without hesitation. But had we got a copy of the piece so that he could play it? We hadn't, so I wrote it out for him in a couple of minutes. There are many able instrumentalists who can only play from notation, and it is closely bound up with how we hear music, and of how we think of music, which is discussed in the final chapter. But initially, this raises the problem of how I should represent pitches, notes, intervals and so on. There are instances, such as in Fig. 4, where standard western musical notation is a convenient way, though even there, it cannot really represent what is intended. Notation is only an attempt to indicate something which pre-existed either as sounds or in people's heads; I assume that to be so.

Regardless of whether readers play instruments, they will be able to run through many tunes 'in their heads'. And despite the vast number of terms musicians have invented, they don't have one corresponding to 'visualise', and the story above makes it obvious why we can't talk about visualising a tune. If one could visualise a tune as notation one could, presumably, play it, and I'm sure many readers who can 'think' tunes, don't. I have used a verb for this process for many years: to *auralise*. I have never found it in any music reference book but I was astonished not to find it in the *Oxford Dictionary* either. Music has far too many terms already and I don't introduce new ones without justification. An *oral* tradition is satisfactory for the transfer of song, for oral means by mouth; the French violinist could vocalise the tune. It would be impossible for much of the traditional instrumental music used for dancing which is always directly transferred by one player hearing another play, to be vocalised. That is an *aural* tradition. The reason for introducing this now is that when pitched sound was discovered and music began, it was entirely aural and it went on being entirely aural for a very long time. And we shall get a better understanding of music and hearing if we start the next chapter by imagining ourselves in the position of the people who discovered it and initially developed it aurally.

1.7 Representing intervals

We need a way of representing interval sensations. We can call them names such as 'fifth', or by note names such as C and G, provided that if I put C or any other note name, it does not mean a specific pitch but a pitch to which that of the following note(s) relate. For the time being they represent the sensations of intervals which are roughly those of the modern scale, until we can sort things further. They will be lowest pitch first, unless otherwise stated. If the reader can auralise the intervals, any pitch within, say, half an octave either side of middle C would do for 'C', or they can be played on an instrument. So CG is the sensation of a fifth as pitches in succession. [CG] means the pitches heard simultaneously. Not everyone can auralise simultaneous pairs, or a chord such as [CEG], even if they play a keyboard or guitar, still less if they play a *monophonic* instrument, which can only produce one note at a time.

It is implicit in our ability to auralise intervals in this way, that pitched music is completely *transposable*. It does not matter what pitch one uses for the first note of a tune, one can carry on auralising it; it may be an infelicitous choice if one wanted to sing it, but we've all done that in our time. The primitives who carried out the first experiments discovered that very significant feature of how we hear music, and very puzzling it proved to be. But for practical music the pitches used are limited to the lower part of our considerable hearing range, and it would be interesting to know why.

The word *note* has more than one meaning, and I shall restrict it to the unit of musical sound that we hear. If I need to refer to the things written on staves, I shall call them *dots*: a time-honoured term amongst jazz musicians (particularly those who cannot read them).

1.8 About hearing

The next eight chapters are concerned with how steady-pitched sound was discovered, what sounds were actually selected by hearing from the sensations of instruments as intervals, scales and eventually for somewhat more complex music. I discuss how the harmonics in the sounds may be related to the sensations. The human voice, timbre and loudness are briefly considered. That raises many questions, to which the answers, if there are any, must lie in what our hearing system does. And in Chapters 9 and 10, which give a much simplified account of how our ears and auditory brain appear to work with music sounds (in so far as we understand them and it can be simplified!), we see to what extent that account provides answers. But it may be helpful to appreciate one or two things about our hearing system while we are accumulating the questions.

Sensation occurs in the most advanced part of the brain called the *cortex*, the last stage of the hearing process. There is much more to the hearing system. Hearing evolved to detect natural sound and to provide animals with the best chance of survival. Ears and the first part of the auditory brain are an automatic system, evolved to detect when there is noise, how big the noise is and the direction the noise is coming from, and to produce automatic reactions to detecting those things simply for survival. We all inherit the machinery which does that, and there is nothing to suggest that it has changed in any way in the past hundred thousand years; it is very like machinery going back several million years, for it is similar in the brains of our animal relatives (see Chapter 12). It evolved using entirely natural noises, transient sounds, and everything which occurs in a cortex as a sound sensation has been through that machinery, designed to produce automatic reactions regardless of whether the creature's cortex could, or nowadays can, observe anything about the sensation beyond 'noise'.

What we call the properties of our hearing, some of which seem peculiar, are actually those things selected to make the automatic system as effective as possible for its natural purposes. That includes the range over which our hearing works, from very low to very high sounds, our ability to assess loudness, and so on. Our automatic survival system determines what we can hear and the form in which the cortex gets it. The cortex can't do anything about it; it has to take the sensations it is sent in that predetermined form. It is unlikely that sounds with steady rates of vibration producing the sensations of musical pitch, were generated more than fifteen thousand years ago. There is no possibility that our hearing machinery was evolved to deal with such steady vibration sound. It is fortuitous that the ears and automatic machinery process that sound in such a way that the cortex gets the sensations as steady pitch with tone; as it happens, that kind of sound is of least use to the survival system. The machinery determines that everyone with normal hearing receives similar sensations from the same musical – and of course other – sounds. When one considers all the opinions and arguments people have about music when we do hear the same sounds, what would life be like if we didn't? Since I have not been entirely successful in keeping my own opinions about music out of this book, perhaps this is the moment to introduce the other major component of music: time.

1.9 Time in music

Of all the factors involved in music, how we accommodate *time* is probably the most difficult to understand. It is an integral part of every transient, since each is characterised by the rate at which it changes with time. Steady pitch only has duration. But the beginning of every note or new sound in music is indicated by a *starting transient*. In music, we are sensitive to when a succession of sounds occurs in time by when each sound begins in relation to when the others do. Can one call that a sensation? For the moment we will use a neutral term such as the time component. The distribution of notes in time is just as much a characteristic of a tune as are the pitch intervals. But underlying the notes there is a time pattern. It may be blatantly obvious, but in more time-sophisticated music it is often difficult to decide whether that pattern is present in the physical sound or whether the listener is creating it as a response to obtaining the sensations. In jazz, latin-american music and Shetland fiddling, for example, a regular time pattern is created, but analysis shows that the transients diverge by small amounts from it. It cannot be indicated in notation, and has to be acquired aurally, but it is not simply imitating and it does not take us any further forward to say that people copy a style. Some players of all kinds use minute timing variations when playing, and some don't. The surprising thing is that some can and some can't. Experience can improve the sense of pitch and the ability to play in regular time, but the ability to manipulate fluctuating time is a complete mystery. Of course, some people are natural players of ball games, some can draw, so likewise only some can manipulate timing. But what we shall find in Chapter 10 is that the small time variations which make all the difference to some kinds of music, are large compared with the minute time differences with which our hearing's automatic machinery operates.

In Chapter 11 we will see how far we can get with these problems about time, but time components are a basic element of music, so fundamental that it is legitimate to call time-patterned sound produced only by percussion instruments 'music'; West African drum music clearly qualifies. And whereas one sound repeated at regular intervals of time can occur in nature, though no one would call it music, sounds in a continuous regularly repeating pattern in time do not normally occur in the natural world either. The most frequent objection I have received to this suggestion is, what about a trotting horse? Well, what about a horse with no metal shoes, galloping in its herd across their natural habitat, steppe grassland? Even the background sound of a televised horse race is artificially added and probably computer-generated.

These two phenomena: *steady pitches and time patterns, both of which are artefacts*, inventions of the human race, appear to *delineate music from all other sounds*. Unlike other art forms, music appears to be entirely artificial. It uses the properties of both the physical and biological world, and we discovered entirely empirically how to use them to produce sensations many people like. That still does not tell us why we like the invention, why some people are attracted to it from the time they first hear it, or why millions are now addicted to hearing some kinds of music for hours on end day after day. It is easier to understand why some people become addicted to foods or drinks, or games, than to playing music, and some even to writing music. Gifted composers used to starve in garrets in order to compose music; perhaps most of them still do, for they are exceptionally rare today. What we want to know is why music happened, because nothing would have happened if we had not been intrigued with it, from its very origins.

2. Aural archaeology

2.1 The origins of music: making noises

Many animals can produce a few noises to communicate; we were originally no different, though we eventually discovered how to develop them into complex language. But other animals do not make noise unless it is for vital purposes, such as proclaiming territory, in association with mating, or for mother–offspring bonding; otherwise, an animal's behaviour is naturally selected 'not to make a noise'. Both as hunter and hunted, and we were both, the animal which makes least noise and has the most acute hearing survives and succeeds (what a fine text for this noise-ridden age).

When we became toolmakers, making noise was an inevitable part of operations such as shaping flint. We don't know when we got the idea that we could make noises using artefacts as a distinct action, but it is unlikely that we did so simply for the sensation of hearing them unless and until we were in a sufficiently advanced self-protecting community that it was not disadvantageous. It pre-supposes either a static community or a well-organised large nomadic one; it is unlikely that even the simplest of noise-making for pleasure occurred earlier than about 15,000 years ago, by which time primitive language was developed (see e.g. Boughley, 1975, Clark, 1969). Early experimenters with noise may have realised practical uses for their discoveries, so that when artefacts are found which could have been used to make noise, we should not automatically assume that they had anything to do with music. An obvious use is to signal. The cacophony which people made to frighten away animals or evil spirits was hardly music either.

Although one assumes that percussive sound was the earliest form of music, making noise randomly communicates nothing, and the idea of making it regularly had to come from somewhere. For reasons given in Chapter 11, it may have originated with the noise of ornaments worn during primitive group movements. It was certainly invented independently in several different cultures, and developed in some to a high level of sophistication. It appears that it continued to be associated with dancing, and one gains an impression that until the widespread mixing of cultures took place over the past two or three centuries, percussion and dancing cultures did not particularly develop the use of steady-pitched sounds, and conversely the cultures developing the use of pitch did not evolve very complex time patterned music. But the rapidity with which the two elements have amalgamated in recent times suggests that the sensations of both elements are widely liked as soon as they are experienced.

2.2 Noise-making artefacts

The starting point of percussion instruments appears to be very simple: we just hit something. Discovering accidentally that something one blows with the mouth makes a noise seems more complicated, but both depend on the same principle. Indeed, until the age of electronics, the number of ways we had discovered of making sounds at all was remarkably small. We hit things, blow things, or pluck and scrape strings, and that is still the basis of virtually every artefact we use to generate sound today. They all depend on the same simple principle. What our ancestors discovered by trial and error, though they didn't know it, was that the only things which make useful percussion instruments are those which vibrate and go on vibrating of their own accord after they have been hit. That vibrates the air in contact with them and produces the sound. If they don't vibrate one either gets sound from the beater or nothing. The same principle of sound production applies to air in an enclosure such as a tube, or to a stretched string. They have a natural way of vibrating if we can persuade them to do so, and produce sound. The way something will vibrate naturally, if suitably provided with energy, is called *resonance*. It is an extremely important phenomenon; amongst other things, our ears use it to detect sound, but we can get by without any technical knowledge of resonance, as long as we know the simple properties and consequences.

With percussion instruments or strings, the energy which initiates vibration and is used up in vibrating the object, is provided with one blow or pluck. The vibration dies away, sometimes very rapidly. It is transient. Hardly any of the energy is used to make the sound; it is almost entirely used up in making the object change shape repeatedly, and how long the vibration and therefore the sound lasts depends on the material. For example, much less energy is used up in making steel bend than in producing a corresponding distortion of wood or gut, and therefore the sound from metal lasts much longer. But after the first fraction of a second when an object is struck, the *rate* of vibration itself is reasonably regular because that is the property of resonances. The common resonant property of anything which makes a good percussion instrument is therefore that it is likely to produce reasonably steady pitch after the initial starting transient, though the size of the vibrations continuously decrease. An alternative way of making something which will vibrate resonantly do so, is to feed it with energy in a form which is vibrating at the same rate as it will naturally vibrate. Sing a note with the sustain pedal of a piano depressed, so that the strings are free to vibrate, and the tiny amount of energy in the vibrating air will cause one of the strings with the same natural rate of vibration to start vibrating by resonance. The principle is used in having sympathetic strings on archaic and folk instruments.

The same principle, that it requires very little energy to make something which is resonant vibrate, is utilised in wind instruments. One can force the air in a tube to vibrate at any reasonable rate, but at a natural resonance of the column of air, it can produce a lot of sound for comparatively little energy; the player feeds in energy with the tube's natural vibrating rate. That is what actually happens, and it is not the simple process one might presume from elementary descriptions. Once the player has started the air in the tube vibrating, the vibrating air itself plays a considerable part in controlling the rate at which the little puffs of air are provided by the player. Players don't know they are providing 440 puffs a second or need to know that; when they begin playing today they discover how to co-operate with the tube to produce the sound by trial and error, just as they did after the first accidental discovery that doing it produced steady-pitched sound.

What is far more difficult to make is a useful percussion instrument that does not have an identifiable steady pitch; it has to resonate to produce sound, and it must have a very large number of small resonances in order that our hearing does not distinguish any of them. That is the behaviour of a good cymbal (it happens also to be that of the box of a violin). It was found that pieces of wood when struck produce pitched sound, and if by trial and error they were shaped to make more noise, it inevitably also made the pitch more obvious, as with the xylophone. Broadly speaking, if artefacts made a usable noise when manipulated, it was almost always reasonably steady-pitched. The special property of devices which produced sound by blowing them was that they produced a *sustained* steady pitch, and once mastered, with steady loudness too. People did not have to recognise that there was anything unusual about the sound; that was the property of an artefact sound

2.3 Sustained-pitch instruments

On the evidence of artefacts, the oldest musical instruments producing sustained steady pitch, by some thousands of years, are the panpipes and flutes. The only naturally occurring objects which without modification can produce a steady-pitch sound by the simple and accidental process of blowing across the end, are pieces of hollow plant tube in which one can excite the resonance of the column of air. For the purposes of this discussion, a panpipe is a tube with one end closed; a flute is a tube with both ends open, whether blown across the hole at the end or across a hole in the side near one closed end. Some little while later, New Stone Age man also discovered the whistle mechanism or fipple, in which air is blown through a narrow slot to impinge on a sharp edge on the far side of a window; it was used to produce resonance either in a tube in the manner of a recorder or in a small pot like the modern referee's whistle, sometimes called a pot-flute. The fipple is a considerable advance in making a sound-producing artefact. It isn't too difficult to make one in wood with a penknife, but making anything like it out of bone with flint tools beggars the imagination. Many pot-flutes were made, but they were never developed for music. Everything suggests that initially people just enjoyed making noises for the pleasure of doing so. Small children of similar mental attainments also do; usually their parents do not.

The panpipes, flutes and whistles were all discovered independently in the first three centres of settled mankind, around the eastern end of the Mediterranean, in Central America and in the Far East, and no doubt elsewhere as well. These together with the xylophone, were the instruments on which the very first stage in the invention of pitched music, the selection of a set of pitches, took place. Tubes excited by vibrating reeds, and by using the lips as a vibrator – the lip-reed process now used for brass instruments – began to appear only some four or five thousand years ago, from which time there are also the first remains of devices which could have been stringed instruments. Contrary to folklore, the hunting bow may have been one of our oldest valuable inventions, but an efficient bow has an inelastic string which does not vibrate when plucked, and it is not really tensioned until an arrow is in the firing position. It is doubtful if it was directly concerned in the origin of stringed instruments, which may well have had precursors in the form of tensioned strips of wood fibres. So far as western music is concerned, bowing strings to produce sustained sounds is believed to have been invented in Asia Minor sometime around AD 900. I do not know whether it was discovered independently in other parts of Asia where it is now used for traditional instruments.

The reaction of the first person in any culture to obtaining sound accidentally by blowing across the end of a panpipe tube was probably fright. Give a young chimpanzee a toy bulb-operated horn to play with and all will be well until it accidentally presses the bulb, and it will jump away in fright. In a world where the sun, rivers and fire were living spirits, it is no wonder that music was magic and has continued to be so, even until today. It is a long process from there to finding that sound-production is a property of tubes, and then by trial and error to connect tube lengths with different sounds. Panpipes were almost certainly one of the first multi-pitched musical instrument producing sustained sound, for a bundle of tubes tied with fibre requires very simple technology. They would produce an arbitrary collection of pitches, but the device made it possible to select pitches. In the Eastern Mediterranean culture, the ends of the tubes were closed with beeswax plugs which might suggest they were 'tuned', but that would conceal the relationship between the internal length of the air-column and the pitch sensation; the sort of sums needed to obtain such a relationship would have been far beyond priestly wisdom then (and in most cases now). The clay-cast panpipes which appeared in the Central American culture suggest that they might have established an empirical set of lengths. The traditional Andean panpipes are still made of wood. Their powerful starting transients would also be effective in producing time patterns for dancing, regardless of their actual pitches. Unlike many archetypal instruments, the sound is quite attractive in small doses when used in a traditional way.

Obtaining two pitches from one tube by opening and closing a hole in the side would also have been an accidental discovery, but it was only developed on flutes and whistles, and not on a panpipe tube closed at the end. A side hole does produce a second pitch, but like a pot flute, it does not behave 'logically', that is to say, the new pitch is not consistently related to where the hole is made. That is explained in Chapter 4, but primitives must have reached that conclusion after many fruitless experiments, just as eventually they found there was a connection between where one made a hole in a flute or whistle barrel and the sound, though it wasn't simple – and it isn't simple. We don't know whether, for example, any primitives had sufficient logic once they were trying out three or four holes in a tube, to space the holes equally, because that would not have taken them on the road to pitched music. They selected holes with a particular kind of spacing, by sensations. In Section 3.9 we discuss a case where some holes were put in by visual rather than aural logic and its consequences.

The unique thing about steady-pitched sounds is that if two of them are heard in succession, and their rates of vibration are related in particular ways, that produces a characteristic sensation, which, it would appear, the majority of the human race likes hearing. The steady-pitched sounds which artefacts can make can have any arbitrary rate of vibration. So the ones which were liked are, in some way, a feature of *people*. Pairs of such sounds may also produce unique sensations when heard simultaneously; but that is a very different matter, as we shall see.

3. Hearing selects intervals

3.1 Introduction

We had seen how to produce steady-pitched sounds. We can now consider how the discovery of the sensations of intervals between such sounds would have come about, and some of the implications. If you are unfamiliar with traditional nomenclature, a sixth is any pair of pitches with the sensation created by CA, a fifth CG, a fourth DG, a major third CE, a minor third EG, and a second DE (and see Appendix 1).

Whether it was with panpipes, flutes, whistles or xylophones, after an initial period when people simply made pitched noises, some experimenters selected devices making pitches which created sensations of the intervals listed above; and these intervals we use in music today. People liked them; that has been the story of music ever since. It may suggest that characteristics of these artificial pairs of sounds interact in an unusual way with a hearing system. Later in this chapter – and perhaps ten thousand years later in terms of development – there is an account of the way in which the physical features of sounds which create these sensations were discovered. But the investigation of instrumental sounds only describes the physical sounds our hearing system selected. It does not describe what we hear, still less why we selected them. So the historical picture is very important.

The aboriginal process started with simple panpipe tubes which produced a random collection of pitch sensations when played, until by accident someone found that two of such tubes produce a pair of pitches which are a 'nice' sensation: more attractive to him than the rest, when heard in succession. It almost certainly was one of the intervals listed above. The primitive had to observe this, connect the sensation with some property of the tubes, and then randomly try other tubes and observe whether some other pairs made a nice sensation. The sensation was ephemeral and elusive. It had to be remembered to be recognised again, and fifths can occur between pairs of high-pitched and of low-pitched notes. Someone with a bundle of crude panpipe tubes gradually selected pairs and threes and fours which produced 'nice' sensations between them.

Carrying out blind experiments by making holes in flute and whistle tubes is far more complicated. A hole in the side of a tube allows the production of two pitches; they may not make a nice sensation. If they do, the next hole made may not make one with either. The processes involved in experiment and attempts to find the secret of the magic were unimaginably time consuming, and could only have been tolerated in well-provided communities. Flutes were eventually made with four finger holes producing 'nice' sensations between its intervals, though archaeologists are not always cautious about interpreting such artefacts. In the Andes today, some flutes are played entirely by cross-fingering, so that the spacing of holes does not necessarily allow one to calculate the pitches the pipes were intended to make.

But one aspect of the prehistoric process of musical discovery is of major significance. One cannot blow two panpipes at the same time, and flutes and whistles only produce one note at a time. The whole selection process was done with *consecutive* pitches. Our ancestors selected pairs of pitches because they liked the sensation of one compared with their *memory* of the other. The use of short-term memory associated with the sensation of pitched sounds was fundamental to the whole process of the way the auditory cortex selected pitches to achieve attractive musical intervals, right from the start.

3.2 The pentatonic scale

The remarkable thing is that instrument-making musicians not only selected intervals with which we are very familiar, but, in geographically isolated cultures in several parts of the planet, they converged on one and the same set of intervals, which became the first widely used set, and which is still used today in several parts of the globe in some of the traditional music which has survived (Yasser, 1975). This is the *pentatonic scale*, which we can represent to start with, by the intervals we hear between the pitches of the notes C, D, E, G and A of the modern scale. This scale was evolved using panpipes, flutes, whistles and, very significantly as we shall see, using xylophones and gongs. From whatever collection of arbitrary pitches on whatever instruments they started, some process was operating in all those early experimental musicians in different isolated communities, which guided them to this, the first widely used scale. One cannot escape the conclusion that what we are concerned with is a common property of human hearing.

And because interest has concentrated upon the intervals between the pitches, one may overlook that it always turns up as five pitches. The pentatonic scale is the starting point from which, later, the seven-note scale used generally in orthodox music was developed, and from that came the twelve-note scale used universally in more advanced western music today.

3.3 The pitch ratios

To get any further we need to know what consistent relationship there is between the actual sounds which always produce the fifth sensation, the fourth sensation and so on. The pursuit of the basis of these intervals goes back to classical Greece if not earlier. The School of Pythagoras used a monochord: a single tensioned string with a movable bridge which would divide it into two lengths. They found that when attractive intervals as judged by hearing were obtained, the lengths into which the string was divided were in simple ratios. If the two vibrating lengths of the string were in the ratio of 1:2 the sounds were, in our nomenclature, an octave, while a ratio of 2:3 produced what we call a fifth. They were, apparently, more intrigued with the magic of the simple arithmetic than the serious musical implications, and we have no idea whether it was related to the music of the time, or what that was really like. Ellis (1885) pointed out that monochords are not accurate devices and suggested that some jumping to simple conclusions occurred. That seems to have gone on ever since in offering physical explanations for music!

By early in the seventeenth century, it was recognised that sound was a vibration. Galileo deduced the relative rates of vibration of musical intervals from studying strings, but the remarkable French philosopher monk Mersenne in 1637 deduced the relationship between length, tension and the actual rate of vibration of strings, using ropes and wires over 30 metres long on which he could count the vibrations. There is more significance in the use of strings for these investigations, discussed in the next chapter, than just the ease of measuring lengths. Unlike some scientists who have investigated musical phenomena, Mersenne was also a very gifted musician and no mean intellect; he was a colleague of the philosopher Descartes.

The era of serious direct measurements really dates from 1819 with the invention of the siren by Cagniard de la Tour. A siren is a large circular disc with a row of many evenly spaced holes around the perimeter, which can be rotated at known speeds. When a jet of air is directed to blow through each hole as it passes, it produces a pitched sound, and the pitch is related to the rate at which the holes go past the jet. It was therefore possible to measure the rates per second when two pitches produced by the siren made octaves, fifths or other interval sensations, and to match the pitch to the pitches of notes of musical instruments by hearing.

The siren confirmed that there were apparently very simple relationships between the rates at which holes passed the jet and two pitches producing one of the interval sensations of the pentatonic scale. For example, whether the rates were 200 and 300, or 300 and 450, or 50 and 75 holes per second: all having the ratio of 2:3, they always produced the fifth sensation. No one appears to be have tried, for example, 303 and 450 holes in succession, to see whether that also produced a fifth sensation. However, the account (Helmholtz, 1870) of how the measurements were made, which was to engage and then disengage a mechanical revolution-counter by hand while watching the second-hand of a clock, is so crude that a little jumping to conclusions about the exactness and simplicity of the relationships still took place.

Just as with the tuning of an orchestra described in Chapter 1, all these experiments are using hearing to assess pitch. Some fifty years later, as we discuss in Chapter 4, it was discovered that the sound of the siren, the cello and of all other musical instruments, consists of several rates of vibration, called the *harmonics*. But we only get a single sensation of pitch from them. So what does

CD DA	EA DG	CE	EG	CA	CD DE GA
5th	4th	major 3rd	minor 3rd	6th	2nd
2:3	3:4	4:5	5:6	34:5	8:9*

Table 1. Approximate ratios of pitch-frequencies in the CDEGA pentatonic scale. *See Section 6.5 and Appendix 2.

the rate at which the holes pass the air jet mean? Only that we can match two sensations by hearing. However, a tuning fork produces only one rate of vibration: a *pure tone*, and if our hearing matches the pitch of a tuning fork vibrating at 200Hz to the pitch of the siren, we can use the siren to ascribe numerical values to pitch. What are we going to call it? We cannot simply call it the frequency of the note. The note has several frequencies, we haven't actually measured any of them directly, and what we have assessed has only the accuracy of our hearing. We will call it the *pitch-frequency*, as a constant reminder that it isn't a physical measurement but a value attached to a sensation. We can now measure the wavelength of light which produces the colour yellow, but it does not explain why the sensation is yellow, or why yellow and blue produce a sensation of green. We hear two pitch-frequencies and get a fifth sensation. We want to discover why. Amongst other reasons for using pitch-frequency, there are circumstances described later, when we can hear a pitch, and put a value on it, and there is no sound with such a rate of vibration entering the ears at all.

All these investigations showed that an octave sensation is always produced when the pitch-frequency of one sound is twice that of the other, and that there are ratios which will always produce the sensations of other intervals such as in Table 1. Those are all the intervals which can be played by selecting pairs of pitches from the pentatonic scale CDEGA; DA is a fifth, EA is a fourth, DE and GA are seconds. So our ears are capable of coding steady-pitched sounds and delivering them to our cortex in such a form that they produce identifiable sensations when the pitch-frequencies of consecutive sounds are related in these ways. The hearing of musicians selected sensations produced by sounds with those ratios more than ten thousand years ago because they liked them, and they and some listeners went on liking sensations with those ratios without knowing they were vibrations in simple ratios, until the present day. In fact the story is not quite so simple.

3.4 Derivation of the pentatonic scale

Suppose we start as our ancestors did, but with the enormous advantage of being able to give names to the notes and intervals, and select successive pitches by hearing. We have a crude whistle, panpipe tubes or xylophone bars of wood, and start with an arbitrary pitch we call C. We discover another pitch we like when we hear it before, and after, sounding the C. It could have been E or G or A, or less probably D, each of which make a simple ratio with C. Since it could be any of them, call it X. We then experiment until we find another pitch which we like in succession with C, and we also like with X. We now have three pitches which sound nice in succession in pairs: three pairs. We find another pitch which sounds nice with each of those three in turn, so that we now have six pairs of pitches, each of which sound nice in succession. And then we find a fifth pitch, which sounds nice with the other four, which makes ten pairs in all. Doing this by trial and error, even with panpipe tubes, is an iterative process of monumental proportions. But everything man discovered, and a great deal of what he still discovers, is suck it and see, or in this case, blow it and hear.

And do we end up with CDEGA with the exact ratios between the pairs given in the previous Section? The process will select five pitches, though not necessarily CDEGA. Before we discuss that, we need to think a little further about pitch-frequencies. Suppose we happened to select the pitches in the order C, G, D, A, E; there is nothing to indicate that we did, though the archetypal Chinese scale is, in western characters, written in that order. Each pair, C:G, G:D, D:A and A:E are fifths – if the sequence spanned more than two octaves. The five pitches of the Chinese scale, when arranged in pitch order are CDEGA within one octave, like other pentatonic scales, because the Chinese notation goes from C up to G, drops from G down to D, up to A, and down to E. It doesn't matter in which order we hear two pitches; if the D is lower than the G, it's a fourth and anyone can recognise the difference between a fourth and a fifth whether they know the names or not.

Now if we thought that pitch-frequency was a physical measurement, we would say: C to G is $2:3$, D from that G is $4:3$ and so on until we get an E. Students who have done their homework will know all about this. It is the beginning of a theoretical exercise called the Cycle of Fifths, the gist of which is that if one started with a pitch-frequency C, and went on deriving the pitch-frequencies of one note from another by exact $2:3$ and $3:4$ ratios, through twelve steps until one got back to the supposed 'C' again, it will not be the same pitch as the C one started with. The idea of such a cycle was meaningless until the twelve-note scale was invented many thousands of years later, and we had names for all twelve notes. And it is a cycle of alternating fifths and fourths.

The E derived by successive intervals of the pentatonic scale would have a ratio of $64:81$ (Appendix 2), if it could be carried out with physical precision. But the ratio $64:80$ is $4:5$, the major third sensation. The difference between the two Es is about a fifth of a modern semitone. There is no way in which the intervals can all have exact simple physical values (see Appendix 3). But we have all heard pentatonic scales and tunes and been happy about the pitches. The answer may be slightly disconcerting but underlies the whole of the pitch phenomenon. What constitutes a satisfactory sensation of an interval from two notes in succession, depends on how precisely human hearing can hear anything, and on the nature of the sounds of the instrument. The sounds of real instruments do not have a physically precise value. Two sounds which produce an acceptable fifth sensation from successive pitches have an approximate $2:3$ ratio. This point will be amplified later.

The pentatonic scale CDEGA was not obtained by arithmetic. What complicates the situation further is that on all monophonic wind and brass instruments, the players can easily adjust the pitches and produce intervals satisfactory to their and a listener's hearing, within the margins that judgement can be made. We can't ask a wind player to play exact ratios. The other question is: if a process of selecting five pitches using that one simple criterion of 'liking' them in consecutive pairs is carried out, are we bound to finish up with CDEGA? Well, yes and no. To take the simplest instance, suppose we had got as far as CDGA. The fifth pitch can be either E or F; both make acceptable ratios in succession with C, D, G, and A. If we select F, we get CDFGA. But that is the same pentatonic scale, only the pitches are in a different order; compare FGACD. Try it on an instrument if you are not convinced. Of course if we start with C, the next pitch need not be D, E, G or A. It could be F, E♭ or A♭, all of which make nice ratios with C. But if the same process is followed (see Appendix 2), one will end up with one of the five versions of the same pentatonic scale, that is to say, pitches related as are any five consecutive pitches from the series CDEGACDEG.

And for one fascinating piece of real evidence which supports this, I am indebted to Dermot Barton (private communication). In the Sabah district of Borneo fifty years ago, and perhaps still, the musicians used five different simple flutes, each producing one of the five different arrangements of the scale listed above.

Note that the sequence CDEGACDEG contains EC called an augmented fifth, nominal ratio 5:8, and ED called a minor seventh, nominal ratio 5:9, but it does not contain a semitone such as EF, a major seventh such as CB or an augmented fourth such as FB.

3.5 Using the scale

One reason for a specific set of pitches was because we liked the sound. The other reason is also important in the general development of music. A solo performer can use any pitches and can improvise with those pitches. There then comes a moment when he wants to do the same thing twice. Unless that is remembered entirely in terms of a sequence of finger movements – what we instrumentalists nowadays call muscular memory, the sequence of sounds has to be remembered. This is a tune! Have you ever de-tuned a guitar to a set of random unrelated pitches and tried to remember them? If we liked the pentatonic pitches in pairs because we could remember the first to compare the second with it, we also have a persuasive basis by which a pentatonic tune could be remembered when pitched music started. There are very memorable pentatonic folksong tunes, and instrumental pentatonic folk music occurs all over the world, their origins lost in time.

3.6 The octave

The obvious pitch-frequency ratio missing from early discoveries is the octave, the simplest of all ratios, 1:2, which so many who discuss the phenomena of music seem to regard as unique or special. It may have been produced with some early panpipes and xylophones, and it would not be surprising if it was dismissed as being rather uninteresting, which it is, compared with the other intervals. But there would have been a jinx about it. One can, fortuitously and unreliably, get a pitch closer to an octave than anything else by blowing a simple flute differently; on developed instruments this is called over-blowing. It requires skill and practice to produce a nice ratio octave on a modern instrument, and if one obtains it on a simple whistle it is rarely pleasant. If one tries over-blowing a panpipe, the pitch is a very poor octave-and-a-fifth higher (it behaves similarly to a clarinet), and one can hear that pitch sometimes from native Bolivian players today when they momentarily blow incorrectly.

Historically, by the time musicians became interested in extending woodwind scales to a higher octave, the bore of the tubes had already begun to be modified somewhat from the simple cylinder, and the scale had been extended to seven pitches, requiring six holes; there is a problem if one adds a seventh hole for an octave, that one only has eight fingers, and little fingers are usually short. Various tricks were discovered for getting a reliable further octave of notes by over-blowing, such as moving the blow-hole of the flute away from the end, or making little holes opened or closed by the thumb near the fipple or the reed, and so on. But an octave hole was added to the simple scale on a few instruments in special circumstances (Section 3.9).

Otherwise, there is nothing odd about the fact that if one has two pitch-frequencies such as 100 and 150 with a ratio of $2:3$ and one multiplies both by two, giving 200 and 300, they will still be in the ratio of $2:3$ an octave higher, and they will still produce a sensation of a fifth. Multiply all the pitch frequencies of a tune by any number and the ratios remain the same. If we multiplied them all by about 9/8, we'd have the same tune, only we'd call it transposing up a tone. Since pitches are related by ratios, transposition is automatic; the hearing mechanism automatically accommodates it.

The commonest use of the octave, which people do without even knowing what they are doing, is often called mixed-voice unison singing, which of course it is not, but simply female voices singing an octave above the male ones. Compared with anything we've discussed so far, unison singing by a group of only males or only females is an advanced activity. Everyone has to 'know' the tune – to have learned it. I have no idea when or where males and females first started singing together, but I would guess that some early people would have had taboos against it. Females do sing an octave above males in communal singing – not that it happens very much today. The question is best put the other way round and is like any other selection of music sounds; what else would they sing that sounds nice and would fit the respective vocal ranges? Have you ever heard singing in consecutive sevenths or ninths? It fitted the voice range of male choirs in churches to sing in consecutive fifths for some centuries; I don't know whether they thought it was nice or a penance, because the established church was ambivalent about the pleasures of music for a very long time. Though the question is more frequently attributed to General Booth of the Salvation Army, it was Charles Wesley in 1740 who first asked why the Devil should have all the nice tunes.

The octave phenomenon is also given peculiar significance by the stage of musical development at which letter names were allocated to pitches, around AD 1000. If literacy had been advanced when the pentatonic scale was discovered, the five pitches might have been labelled A to E (and if there were five pentatonic flutes, even worse mayhem would have resulted). Pitches only needed names when they had to be identified; they are associated with visual representation and a keyboard. When the keyboard was extended, repeating the letters A to G over again, this simply says that notes with the same letter name have the particular pitch-frequency ratio 1:2. And with an extended pentatonic scale CDEGACDEG, a monophonic tune can cross a boundary between the repeats of the letters, but hearing is unaware of this and treats all sensations of ratios in the same way. The lettering system is at cross-purposes with the interval system, as we saw in Section 1.1, because of the nature of the seven-pitch scale which it labels.

3.7 The extension of the pentatonic scale

It is even more doubtful if further pitches were added to the pentatonic scale using the cycle of fifths and fourths, than that the cycle produced the pentatonic scale itself. The idea is clever with hindsight, but although in principle the series F-C-G-D-A-E-B rearranged in pitch order produces the standard white-note scale, the same argument about selecting the pentatonic scale applies. Alternatively, as I have heard with dismay, look at the C-D-E-G-A-C sequence; there's a bigger gap between E and G, and between A and C than there is between the others! If the only pitches you have ever heard are CDEGAC, you have no means by using hearing of assessing that EG or AC are bigger gaps. We only know that they are bigger gaps because we have heard pitches between the gaps. A pentatonic pipe will have four unevenly spaced holes whichever of the five versions of the scale it produces. A row of panpipes with lengths that do reasonably match the pitches CDEGAC, does not provide any logical visual pattern (Figs 1 and 3A). Section 3.9 gives examples of what happened when people did extend the pentatonic scale by visual logic.

There is every reason to suppose that our ancestors discovered and added two further pitches, such as F and B, to the pentatonic scale empirically, just as that scale was evolved, by testing a whole series of nice consecutive sensations, but it involved a decision. There are no further pitches which sound nice in succession with all five pentatonic pitches and with each other. In the scale CDEFGABC, F and B each produced attractive sensations in sequence with most of the pentatonic pitches CDEGAC. But EF and BC do not do so. They are a new kind of sensation, acceptable but with no magic. Their ratio is something like 15:16. The interval CB (8:15) is even less attractive. We can learn to recognise a semitone such as EF. There is a reason for the difficulty (and a particularly

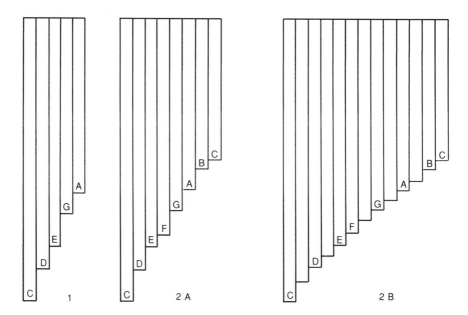

Fig. 1. Panpipe tubes producing a pentatonic scale. Fig. 2. Panpipe tubes for advanced scales. A: Tubes producing a major heptatonic scale, based on the measured internal lengths of a set of Peruvian pipes. B: A theoretical set which would produce an equal-tempered scale.

good reason why most people don't like the two pitches of a semitone simultaneously). And if we look at eight-note panpipes, the lengths of the tubes still do not make a logical pattern (Fig. 2A), and nor do the positions of the added two holes on a flute.

However, there is another consequence of the notes selected by hearing to expand the pentatonic scale. We can add an F, making acceptable sensations after each pitch except E, and a B which does so after each pitch except C, but F does not with B. There is no simple ratio to which the FB interval approximates. It is sometimes called the augmented fourth, and sometimes the diminished fifth, and amateur singers and players find it a difficult interval to produce. To many people the augmented fourth is not a nice sensation, and it was called the Devil's Fourth. My memory is that the tap note and hum note of the bell in the Wills Tower of Bristol University make such an interval, and must be somewhat ominous to those taking examinations.

There are four possible pitches which can be added to the CDEGAC scale, in our terminology, F, F♯, B♭ or B. Each of them on its own makes attractive intervals in succession with most of the pentatonic pitches, except the semitone it creates. But every pair carries the additional penalty of making an augmented fourth. F and B make it. Add F and then B♭. FB♭ is a nice fourth giving a scale CDEFGAB♭C, but EB♭ is the nasty fourth. Add B and then F♯. F♯B is a nice fourth giving the scale CDEF♯GABC, but CF♯ is the nasty fourth. Because we have note names and can look at these scales, the process has produced what we call the major scales of C, F and G. How do we know? Because that is what an elementary tutor calls them. If we have no letters or concept of F♯ and B♭, we hear identical scales, but they don't have identical pitches, and if the Cs match in pitch, the order of the intervals are different too. This would have been very puzzling to anyone who had no conception of what was controlling the selection process and what resulted. If anyone who is conventionally taught today, wanted to know the logic of scales, they would find it equally puzzling, but surprisingly few do. They just accept the scales.

It was a bold decision about pitch relationships which gave rise to the sequence of pitch-frequencies CDEFGABC. It is not unreasonable to call it the *heptatonic scale* because it is the most commonly used set of seven pitches, and dignified by being called major, for unknown reasons. It is also called the diatonic scale which is pure mystique. The *Harvard Dictionary of Music* says that diatonicism is used to describe music in which the tonality is primarily diatonic, the sort of statement which makes one feel there is a secret circle to which one has not been admitted – the fellowship of circular arguments. I venture it would tax most musicians to produce a definition of tonality and, perhaps wisely, most reference books don't try. But it is a real phenomenon; the pentatonic scale does not have it.

3.8 The heptatonic scales

I have written the sections above purposely in the hope that it will persuade readers to think about intervals and scales as products of our hearing system – which they are. They were not necessarily only selected by people using bone flutes and panpipes, but they were selected empirically using relatively primitive instruments and consecutive intervals. However, the recent discovery of a set of bone flutes in China, dating from about nine thousand years ago includes one intact instrument producing a close approximation to a major heptatonic scale (Zhang, 1999). It suggests inventors had developed music a long way on such crude devices by then. But the significant thing about adding F and B to the pentatonic scale was the vastly increased number of attractive intervals it made available, and which, as I will discuss in Chapters 6 and 7, produced the potential for a phenomenal development in music during the recent five or so centuries. Two further pieces of historical evidence are related to this.

3.9 A visually determined scale

In Section 3.7 above I said that if all one has is a pentatonic scale, there is no way in which hearing could inform one that there is a bigger jump in pitch-frequency from E to G than there is from C to D, D to E and G to A. Bagpipe chanters are of special interest because bagpipes are of great antiquity and have

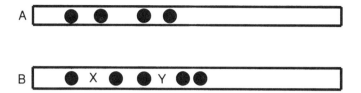

Fig. 3. Finger-hole positions on pipes. A: On an end-blown flute producing CDEGA. B: On a bagpipe chanter producing a pentatonic CDFGB♭C. The Persian chanter has holes at X and Y, equidistant between the DF and GB♭ holes; the Scottish chanter is similar, with an additional hole giving a scale of B♭CDXFGYB♭C.

maintained a traditional tuning for a very long time. A number of Arab and Persian bagpipe scales were measured by Land, and data are given in the appendices of Helmholtz (1870).

The bagpipe uses a drone, a continuously sounded low-pitched note, discussed in the following Section. The players' practical problem with bagpipes was solved similarly and almost certainly independently, in a number of cultures. One can blow two whistles or reed pipes simultaneously; it is very probable that the double-barrelled Greek *aulos* was a pair of reed pipes, and even earlier Egyptian art shows what might be two pipes played in that fashion. In 1949 one Cambridge scientist's party-piece was to play two soprano recorders at the same time. But unless one can carry out circular breathing, and the only instrument for which that appears obligatory is the Australian didjeridoo, a constantly sounding drone has to be supplied from a constant source of air. Traditionally the air is blown into a bag; more hygienically it is supplied by bellows or, as illustrated in an Egyptian terracotta, by a boy alongside the player blowing through a tube (Buchner, 1956). But if the mouth is blowing into a bag, and the chanter is blown mechanically, it cannot be over-blown and therefore no octave is possible. A number of bagpipe chanters do have a hole giving a pitch an octave above that of the chanter with all holes closed. We might deduce that the inventors were already familiar with obtaining the over-blown octave, wanted it, and found that the only way they could get it was by an additional hole.

Land's data enables one to work out that the pentatonic scale of the Arab and Persian bagpipes, before they added holes to make a heptatonic scale, was in the form CDFGB♭C, a pentatonic form which distributes the holes more or less symmetrically with two larger gaps, DF and GB♭. Suppose we mark those gaps X and Y so that the heptatonic scale produced would sound CDXFGYB♭C (Fig. 3). Calculation of the pitch-frequencies of the notes corresponding to X and Y on the Persian chanters shows that the holes produced, in modern terms, a pitch roughly a quarter tone below an E and one a quarter tone below an A. In other words they were placed visually at the mid-points of the D to F gap and of the G to B♭ gap. The pitch-frequencies of notes X and Y bear no simple ratio to any of the six pentatonic pitches. It would be interesting to know the drone pitch used with this scale, but Land did not give it.

Ellis (in *Grove*, 1927) gives values for four Scottish Highland bagpipe chanters, which correspond closely to the Persian scale, but with one further added note at the bottom, so that the scale transposed to our nomenclature is B♭CDXFGYB♭C. The B♭s and Cs are reasonably good octaves; the intervals which form a pentatonic scale are reasonable matches to simple ratios on all the instruments; X and Y are positioned like the Persian chanter. Ellis refers to the scale as either 'accidental' or 'barbarous'; I think 'misconceived' would be a better term. The theory that the scale was brought back from the East during the Crusades is plausible, but in view of the rather obvious way the X and Y pitches were obtained, it is just as likely that two groups of people used the same visual method. The low B♭ appears to be a Scottish invention.

3.10 Drones

The continuously sounded drone is a feature of many genuine folk instruments. Usually, if the melodic pitches are a pentatonic scale of the CDEGA form, the drone is a lower octave C. Such a drone pitch will make acceptable intervals with each of the five pentatonic pitches, and one may ask whether adopting such a drone influenced the selection of the two pitches added to make the heptatonic bagpipe scale, for a C drone would make an acceptable interval with the added F, but would not match happily if a B had been chosen in order to make simple intervals with D and G. This perhaps throws a different light on the traditional Scottish bagpipe scale. It has a C drone, but a B♭ making a 5 : 9 ratio with C, rather than having a B on the chanter. Its orthodox pitches, C, D, F, G and B♭ belong to a heptatonic F scale, rather than to a C scale, for which hearing would select an E and A. Hearing cannot relate the two aberrant pitches to a specific C or F scale (see Chapter 6). In so far as it is possible to say that the sounds are in a key, it is more in F than C. If, however, you have had the misfortune to hear *Scotland the Brave* arranged for bagpipes with military band, you may have reservations about that suggestion, even though the pitches of massed bagpipes are almost as wide as the fairways at St Andrews.

Similarly related drones occur on various primitive dulcimers, and one is incorporated in the hurdy-gurdy, on which the rotation of a rosin-coated pear-wood wheel continuously sounds a drone string – wonderfully imitated in a Schubert song.

3.11 The historical puzzle

The evidence supports the belief that the sets of pitches making the two basic scales, pentatonic and heptatonic, were selected by hearing, to a large extent on woodwind instruments which can only play one note at a time. All the material for the development of music with simultaneous pitched notes, which is an integral feature of *polyphonic* music, appears to be present from the moment the heptatonic scale was devised, if not before. The very substance of baroque and early classical music, and of much else which followed in the late classical and romantic periods, is based upon the intervals of the major and minor thirds, the fourth, fifth and sixth used simultaneously. One would obviously expect that since these intervals produce attractive sensations as successive pitches, they would therefore have been attractive when heard simultaneously, and indeed as we hear them today they are. Why did it take such an enormous amount of time before the sensations were so utilised? We can consider various possible hindrances.

If two or more people play wind instruments together, they must have a common pitch standard. Considering the number of man-years which were invested in discovering the panpipe, whistle, flute and reed instruments and the pentatonic and heptatonic scales, the making of instruments with similar pitches seems comparatively simple. It appears to have been a long time before a standard pitch was adopted even on a local scale. A widely adopted pitch standard is very recent. The mass of data about historical pitch standards suggesting that 'A' in Europe has varied between about 395 and 470 vibrations a second, means comparatively little. Many of the values obtained from old organ pipes only indicate that organs varied a great deal in their 'A' pitch, and organs still vary because they need only to be self-consistent like panpipes and bagpipes. Remarkably, by very many centuries, the Chinese were the first people to establish a pitch standard in the form of a bamboo pipe of exactly specified dimensions. It was surrounded by mystical and political significance; one does not know whether it was used as a standard. Chinese music was apparently monophonic or with a drone, for a very long time; the traditional material seems largely to remain so.

All stringed instruments of the harp or lyre type are potentially polyphonic, and they could be self-consistent in pitches. Initially they had several limitations. A plucked string produces little useful sound unless it is reasonably uniform along its length and is disposed in such a way as to cause a resonant sounding board to vibrate. The highly improbable Egyptian pillarless harp could hardly have sustained any reasonable tension. A friction-held winding peg was widely used, but it requires advanced technology to achieve one which allows accurate adjustment of pitch. One virtue of wind instruments is that they always make approximately the required pitches, whereas strings have to be tuned every time they are used; to what? Every child's elementary tutor starts 'get an adult to help you' and not every adult can do it, even with the aid of a keyboard. In 1997 I advised a Cambridge professor of classics who made a model of a Greek lyre, based on the form of a tortoise shell artefact which had been found with holes in places which indicated it had been part of the resonator of a lyre, the supposed aristocrat of Greek instruments. I privately formed the opinion that it was virtually impossible to tune the strings with any degree of accuracy, and that if it made pitched notes they might sound like a heavily damped banjo. I venture that for a long time, stringed instruments were little more than strummed percussion instruments, pointing the metre of bards; one wonders whether it was entirely David's harp that charmed Saul.

Did the problem of developing polyphony lie in co-ordinating the actions of people? Were groups of instruments played together in the ancient world? Baines (1961) commenting on an ancient Egyptian frieze depicting what he describes as an orchestra, writes 'I wonder what it sounded like'; I would rather not. There are several problems in people playing together involving time as well as pitches, and, as we see in Chapter 11, there are also reasons why a single monophonic player is often preferred for producing music for dancing. And although playing instruments together was greatly facilitated by notation and education, that is not essential. Musically illiterate jazz players can play together, and the gypsy bands of Eastern Europe appear to have developed polyphony through an aural tradition.

From fragmentary material, it has been conjectured that in Classical Greece, music was monophonic and perhaps limited to a selection of notes from a heptatonic scale within the compass of a human voice (Stroux, 1999). An excess of imagination seems to have been used by authors such as Apel (1946) about music amongst the largely illiterate adherents of the early Christian church. Anything which might be described as western polyphony, which made use of the pitches of the heptatonic scales as *simultaneous* sounds, appears to have been developed either vocally, or using organs, within the last one thousand years. It began to be developed in a significant fashion only about five hundred years ago, and the instruments which played the most important role in this, until at least the end of the eighteenth century, were keyboard instruments and the bowed string. It is usually accepted that bowing was invented around AD 900 (Bachmann, 1969). The two types of instrument represent extremes. The keyboard player could predetermine the pitches; the bowed-string players, unless using frets on viols, could play any pitch. But both determined the pitches entirely by hearing. Why was the potential for polyphony not exploited earlier with wind instruments? It may be that the discovery of harmonics by Helmholtz provides some answers, in the next chapter.

3.12 Non-harmonic scales

Helmholtz collected the pitches of many scales from Asia, and there have been several collections since. Some of them contain some pitch-frequencies which approximate to simple ratios; in some one can see no such relationships. Helmholtz believed that the entire basis of European music was the simple ratios of the frequencies of harmonics, which we discuss in the next chapter, so that his calling them non-harmonic is very understandable. Since the evidence appears incontrovertible that the pentatonic and heptatonic scales based on simple-ratio pitch-frequencies arose independently in several places, and probably at different times, there is every reason to believe that other musical systems could have arisen independently, with a different basis. Music isn't the only art which has multiple origins, but all art depends on one thing, whether people like it, and music is what sensations the individual listener likes. That is no criterion for analysis.

Thus, if there is one characteristic of much oriental music which is inescapable to anyone familiar with western music, it is that the pitch of the individual notes is not steady. It is often produced from low tension metal strings, which lend themselves to making such sounds, and singers produce similar varying sounds. No one knows whether voice or instrument originated such music, a chicken-and-egg situation, though there is an argument touched upon in Chapter 5 and discussed further in the final chapter, against the belief that this music originated with the voice too. What would appear to follow from the preceding discussion and seems amply justified in the following chapters, is that varying pitch sounds cannot lend themselves to the development of polyphonic music in the western sense; two simultaneous unsteady pitches are not attractive to most people (the portamento notes of jazz violins and pedal steel guitars is a different matter; they continue to make simple intervals as they change). A lot of non-harmonic music has a drone as well as a melodic line, just as happened in pre-polyphonic western music.

There are some rather more awkward questions we could ask about some of the published data, such as how anyone measures the pitch-frequency of a plucked slack metal string, or of woodwind instruments on which players can easily vary the pitch by how they blow, to one hundredth of a semitone, and we return to this point in a more general form in Chapter 6. But Helmholtz made a splendid comment on non-harmonic scales. He declared that 'Harmony was a European discovery of a few centuries back, and it has not penetrated beyond Europe and its colonies.' One can almost hear the Prelude to Act III of *Lohengrin* heralding the announcement. He wasn't quite right, but the real significance of his statement is that in the nineteenth century, the 'colonies' included most of Africa together with the USA and the West Indies, which had a large illiterate population of African origin, and when they did experience Helmholtz's 'harmony', that is to say, polyphonic tonal music based upon simple pitch-frequency ratios, they adopted it with open arms. They acquired it entirely aurally. And the process has continued, for western-style pop music, primarily using simple heptatonic scale intervals, has become widely popular wherever it has been introduced, and is also now being imitated by local composers in many parts of Asia and the Far East. Through modern communications, music of other kinds and cultures is readily available, and some people find some of it attractive. But music based on the western heptatonic scale and western harmony does seem to have immediate appeal to peoples of many origins, traditions and musical cultures, and it is also being absorbed into traditional music everywhere. Just as with the pentatonic scale, the process does seem to be associated with the one thing they all have in common, the human hearing system.

4. The beguiling harmonic theory

4.1 Helmholtz Resonators

Whatever theories and ideas about the nature of music had been put forward before, the work of Hermann Helmholtz was a major landmark. He demonstrated how to analyse steady-pitched sound into its components, the *harmonics*, and based on that, put forward what may be called a general theory of tonal music in his great book *On the Sensations of Tone* (1870). As we shall see in this chapter, it had and indeed still has a profound effect on ideas about music, both valuable and in important respects misleading. Like many other terms used by musicians, the word harmonic has at least three different meanings, but we shall use it only for a component of a sound, and qualify the word if it is used in one of its other meanings.

A column of air in a narrow tube, such as in a panpipe, flute or whistle, resonates. It produces a pitched note which can be seen to be related in some way to the length of the tube. A volume of air still has a resonance when it is enclosed in a container of any other shape; if one blows across the mouth of a bottle, the contained air resonates and a pitched note is produced. A large pitcher is used in this way to provide a single repeated bass note in the simple folk music of some parts of southern Europe. The pitch of the note depends on the size and shape of the volume of contained air, and it was used in the early pot flutes, exciting the air resonance with a whistle mouthpiece. If finger holes are made in the container, the pitch rises each time a hole is uncovered. This must have been discovered in primitive experiments, for it is a very easy way of getting several pitches from an artefact, but why was it never developed into a serious musical instrument? The ocarina, which works on this principle, has remained as either a crude whistle, a bird decoy or a toy. The puzzle is that, unlike tubes, the effect of opening holes in the container does not have any simple logic. If there are half a dozen similar-sized finger holes, the pitch rises by about the same amount whichever one is opened. Open two holes and the pitch rises again, but the pitches are not readily related. One cannot blame primitives for being confused, because two eminent Victorian physicists could only produce very complex empirical (and different) formulae which approximate to the relationship between the volume of air, the sizes and number of holes, and the sound frequency. That the early musicians did not persevere with the ocarina supports the view that they concentrated on devices which did behave in what we would call a logical fashion. The panpipe also remained in its primitive form, because it behaves as a mixture of air in a tube and air enclosed in a space, and if one makes holes in the side of the tube, it does not behave in a logical fashion either.

The other feature of the sound produced by blowing across the neck of a bottle, is that the sensation is weak and uninteresting, but the very reason for the nature of that sound made it extremely valuable as a scientific instrument before the days of electronics. For Helmholtz discovered that if the enclosure was a spherical globe, and there was only one large opening of a particular size, the air would resonate at only one rate of vibration. He could check that this was so by using the single-frequency sound of tuning forks. The genius of Helmholtz was that he realised the potential of his discovery. He could add a tiny listening hole to a globe, and when a musical instrument played a note, if he heard the air in the globe resonate, that pure tone was present in the sound. A collection of such globes, each resonating at a different pure tone rate, would enable him to discover what pure tones were present in any sound. Each globe was calibrated by checking it against one of his very large set of tuning forks. The globes are called Helmholtz resonators. These experiments were the basis of his major discovery that a pitched musical sound is made up of a number of pure tone components called its *harmonics*. The process is called harmonic analysis. It is also called Fourier analysis, because some forty years earlier the French mathematician Fourier had shown theoretically that any complex vibration could be so analysed into component harmonics.

Helmholtz believed that the nature of the harmonics was the fundamental reason for our selecting and liking musical intervals which have simple ratios, such as we find in the pentatonic scale, and for our liking orthodox harmony, and he developed a comprehensive theory of music based on his discoveries. It occupies a large part of his book. The entire thesis is very persuasive; it has been the most influential explanation of musical phenomena which has ever been propounded. The ideas are relatively easy to grasp, and they have become the standard explanations embedded in the literature, and handed on through generations of teaching. Hence, one of the most important things we pre-sumably need to investigate, is how our hearing machinery uses the harmonic components of musical sound in the ways Helmholtz proposed, to achieve all the things that the standard explanations of music attribute to them.

Harmonics are indeed important in how our hearing system works. In the middle of the twentieth century there was a major discovery which showed that our ears separate the harmonic components of sounds, and if that and subsequent discoveries about the machinery based upon it had not been made, this book would not have been written. So we need to know a little about harmonics. But I'm afraid *Helmholtz's deductions about the role of harmonics in music were wrong*; perhaps back to front would be a kinder way of putting it. I'm aware that no one has ever been thanked for demolishing any simple belief that people find satisfying, especially if nothing is put in its place, so I will do my best. We need to sort out where Helmholtz went wrong, so that we shall know more precisely what explanations we should seek from our hearing machinery.

4.2 Instrument harmonics

The exciting thing that Helmholtz discovered when he analysed the steady-pitched notes of instruments into their pure tone components, was that the rates of vibration of the harmonics were very simply related. Helmholtz showed that the slowest rate of vibration present in the sound corresponded with the pitch-frequency, as demonstrated by the siren (Section 3.3). And if, for example, an instrument's note had a pitch-frequency of 100 vibrations per second, a pure tone harmonic with a frequency of 100Hz was present, and the other harmonics present had rates of vibration of some of 200, 300, 400, 500, 600Hz and so on, in other words, at rates of the pitch-frequency multiplied by simple numbers such as by 2, 3, 4, 5 and 6. Generalising, he said that steady-pitched sounds comprise a pure tone corresponding to pitch-frequency which we call the *fundamental*, and a series of pure tones which are simple multiples of that frequency. It is a useful shorthand to call these the first, second, third and so on harmonic, and thus in the above example we would call the one of 400Hz, the fourth harmonic – but there will be confusion unless we use these names systematically. The fundamental must be called the first harmonic; one could almost predict that somebody would call the second harmonic the first overtone, the third harmonic the second overtone and get all the names one out from the numbers by which the fundamental is multiplied. The harmonic with a rate of the fundamental multiplied by two is the second harmonic. We stick to the logical names even if one or more members of the series appears to be missing from the sound, that is to say, if analysis shows that a steady sound has harmonics with rates of vibration of 100, 200, 400 and 600Hz, we say that it has the first, second, fourth and sixth harmonics and that the third and fifth harmonics are missing.

If we write in standard modern notation, the pitches which approximate to the sounds of pure tones with the rates of vibration of, for example, the harmonics which such analysis shows to be present in the sound of a bowed open cello C string, we get what is called the *harmonic series* (Fig. 4). One can see that the spacing of those pitches approximate to the simply related intervals which our ancestors selected in creating the pentatonic scales: they are an octave, fifth, fourth, major third, minor third and so on. Of course they are; the ratios of the pairs of harmonics are the same as the ratios of the various intervals, $1:2:3:4:5$ and so on. There is an apparent problem over the seventh harmonic, which has caused a lot of argument, but we will look at that at the end of this chapter. It is a very simple step from there to the idea that if two notes an octave apart in pitch are played simultaneously, and each contains a complete series of harmonics, the fundamental of the higher one corresponds with the second harmonic of the lower one, and all the other harmonics of the lower one will correspond with harmonics of the higher one (Fig. 4). The argument can be extended to two simultaneous notes a fifth apart, because the second harmonic of the higher note will have the same rate of vibration as the third harmonic of the lower note. And similarly, one could argue that there will be some matching higher harmonics between any two simultaneous notes

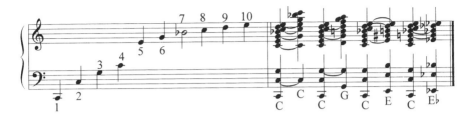

Fig. 4. The harmonic series. Approximate pitches in modern notation of the first ten members of the harmonic series, for example, of a bowed string with pitch-frequency C. The seventh lies between B and B♭ The harmonics of notes comprising an octave, a fifth, major third and minor third; the slurs indicate which harmonics are common to notes of these intervals.

which make any of the simple intervals which were selected to form the pentatonic and heptatonic scales. The explanation is logical, elegant and persuasive. It shows how the simple intervals could readily be related, and an extensive theory of orthodox music can be developed from this. That is what Helmholtz did. One may also argue, as he did, that since all the harmonics of any note of an instrument are related in this beautifully simple pattern of whole-number multiples of the fundamental, that is why we like the sounds.

The theory appears to obtain support from the things which musicians also call 'harmonics'. If one bows an open cello C string while touching the string lightly, halfway along the string from the bridge, then one third of the way, one quarter of the way, one fifth of the way and so on, the pitches of the notes will be quite close to those in Fig. 4. A similarly related set of pitches roughly matching the harmonic series can also be obtained by exciting the air column in the tube of a modern brass instrument to resonate by vibrating it with the lips. The air vibrates as a column the length of the tube, or in two half-lengths which sound about an octave higher, three one-third lengths producing an interval about a fifth above that and so on, and these too are called 'harmonics'. (On some brass instruments one cannot obtain the lowest of the series.) The early unvalved instruments of the trumpet and horn family were ostensibly limited to such notes, and the other notes which make up the twelve–pitch scale became available when the valve mechanism was invented and allowed the introduction of additional lengths of tubing. Indeed it has even been suggested that since those are the actual pitches one can obtain from unvalved brass instruments, that is where our ancestors first heard the simple intervals and adopted them to form the first scales.

4.3 The flaw in the harmonic theory

So what's wrong with the Helmholtz theory of music? Well, in the first place, the pitches of the pentatonic scale were selected by trial and error from consecutive pitches on monophonic instruments like panpipes, flutes and whistles, and it was thousands of years before any notes were played simultaneously such that matching harmonics could be compared in the fashion suggested by Helmholtz. Pitch is a single sensation, regardless of the instrument producing it. We remember the pitch of a note and compare the memory of it with the pitch of another note. That is what actually happened. If we selected consecutive pitches by Helmholtz's theory we would have to remember the third or the fifth harmonic of the first note and match our memory of that harmonic with a suitable harmonic of the second note. That is unconvincing and it is really pure speculation.

Helmholtz belonged to the grand Victorian age of elegant theories, when everything one discovered in science suggested that the entire universe operated to a beautiful plan. One would never accuse Helmholtz of fudging a result, but the truth is that his globes could not measure harmonics with very great accuracy. Even a narrow resonance starts vibrating a few vibrations per second below its maximum rate and goes on doing so for a few vibrations per second above it, and the resonances of the glass globes were not that sharp. But in the age of Helmholtz, if measurements did not exactly fit a beautifully simple theory, you assumed that the problem lay with measuring, and that there was a beautiful plan. There are dozens of examples of this in nineteenth-century science. Interesting as are the measurements of a hundred different ethnic scales in the appendices of his book, I'm afraid they all have to be taken with a pinch of plus or minus one per cent (and several cents, Appendix 6).

But there is a far more serious assumption, which has also been made by people later than Helmholtz. If the sounds of modern instruments are analysed they do have harmonics which are quite close to an ideal series, close enough for Helmholtz to have jumped to his conclusion that most instruments produce a complete series of simple multiples of the fundamental, while the clarinet produces the odd numbered multiples, 1, 3, 5, 7 and so on. And there is a simple story, with simple diagrams, which most musicians are told if they are told anything. A tube open at both ends 'like a flute' produces the complete series. A tube closed at one end 'like a clarinet' produces only the odd numbers. The next time someone says that, ask how oboes, bassoons and brass instruments manage to be open at both ends, and it is only the clarinet which has one end closed.

The truth is that the simplest cylindrical tube open at both ends, the end-blown flute, produces sound with a series of harmonics, but they are not exact multiples. The modern flute does reasonably well, by having the aperture in the headpiece away from the end of the tube, and by tapering to a narrower bore between the headpiece and the main barrel. The simplest tube closed at one end, the panpipe, does produce only the odd number harmonics, but they are not a nice series either. And, as explained in modern books such as by Benade (1976) on the physics of instruments, all the other instruments would behave like a very bad clarinet if they were arbitrary tubes. Both the specific diameters and the tapers of the tubes of each instrument have been very carefully selected to cause them to produce the sounds we associate with them, when played by a skilled player. The best modern plastic recorders are exact copies of the best traditional recorders, and their sounds and behaviour are remarkably like the best traditional instruments (Unwin, private communication), but they are not a simple shape.

It is doubtful if there has been a significant change in the construction of orthodox monophonic instruments since the time of Helmholtz. He did his measurements on evolved modern instruments, and they do have sounds with harmonic series reasonably close to the theoretical series he described. If Helmholtz had made measurements on wind and brass instruments three hundred years earlier, let alone three thousand, it is doubtful that he would have discovered anything on which to erect any theory that intervals, scales or harmony are based on harmonics. So what did he actually discover? That sounds of the kind he analysed were the end-point of an enormous amount of painstaking development carried out by instrument makers during the eighteenth and first half of the nineteenth centuries. They had started with things like the shawm, which Macgillivray (in Baines, 1961) described as having a 'jangle of overtones' and finished with the modern oboe. But they were working entirely empirically by trial and rejection, using only what their hearing and that of their clients liked, and what Helmholtz discovered was that their hearing liked sound with harmonics reasonably close to the theoretical simple whole-number series. It was hearing which selected that, without knowing what it was selecting. But the harmonics of wind and brass instruments were nothing like a set of simple whole-number multiples five hundred, let alone ten thousand years ago when intervals were selected. This also provides a reason why primitive instruments produce what to us are such unusual sounds.

As anyone with a good sense of pitch can readily detect, the touch 'harmonics' of a modern cello string all sound 'flat', and indeed they are all slower rates of vibration than theoretical expectation; whereas, as measured by modern electronic methods, the true harmonics of the sound of a normally bowed steel-cored string are all 'sharp' with faster rates of vibration than theoretical (Beament, 1997). Modern string-makers subject strings to extensive physical tests and have a useful term, *harmonicity*, for how closely the harmonics of a string's vibrations do approach the theoretical exact whole-number multiples (Pickering, 1989). But even the harmonics of a bowed rough-twisted gut string like the ones used three or four hundred years ago, are much closer to the theoretical series of simple multiples of a fundamental than any of the woodwind instruments of that period.

As for the brass instruments whose 'harmonics' were suggested as a source of the simple intervals, the original horns were actually animal horns with the narrow end cut off to make a crudely tapering tube of arbitrary shape. The virtue of horns, if such it can be called, is that they were capable of making a very loud noise. The description applied to the sound of those used by Jews for ceremonial occasions is 'awesome'; if they didn't actually make the walls of Jericho fall down they would have terrified the guards. If the sound of the lowest obtainable note was awesome, the pitch relationships of the 'harmonic' series obtained from them would be no less to any musically sensitive ear. A huge amount of trial-and-rejection experimentation was carried out to try to get the brass instruments to produce a complete set of 'harmonics' which were reasonably close to the pitches of a set of simple musical intervals, using hearing, which did of course also make the sounds more acceptable, but one can judge how successful it has been by listening to the bugles of any school cadet band (the players are allowing the tubes to determine the pitches because they have not learned to modify the pitches with their lips, as brass players must).

Indeed, because of the resurgence of interest in early music, it is often now played on reasonably authentic reproductions of early instruments, which have been made as a result of extensive research. One can hear what instruments of a few hundred years ago might have sounded like, though even they represent a considerable development from the sounds with which our ancestors selected the pentatonic scale. Unfortunately one's all too frequent experience is of the playing of rackets and cornetts, with the odd sackbutt and serpent, by entertainers who rely on the listeners to be more intrigued with the quaintness of the devices than the pitches or the noises, and I am not alone in reacting on such occasions with a mixture of amusement, disbelief and agony. An instrument like the cornett is capable of making interesting if rather unusual sounds in the hands of experts, which indicates that alongside the substantial improvement in the harmonicity of instrument sounds developed from these forerunners, the instrument makers were concerned to make them more user-friendly too. If players have to practise for years to produce acceptable sounds on modern instruments, the same must apply *a fortiori* to earlier ones.

But what gives the Helmholtz harmonic theories the *coup de grâce* are xylophones, gongs and bells. Pentatonic scales were certainly selected on xylophones and their even older stone equivalents the lithophones, and the Chinese did so with their strange L-shaped metal plates and gongs. For example, the rates of vibration of the harmonics of a typical modern xylophone are the pitch-frequency multiplied by 2.76, 5.4, 8.9 and 13.3, while gongs and bells are no less irregular. The harmonics of a kettledrum have been measured as 1.61, 2.19, 2.30 and 2.54 times that of its fundamental, but they provide a single pitch. The xylophone was about the only type of primitive instrument on which a player could sound two pitches simultaneously, and a lot of help its harmonics would have been towards selecting simple ratio pitch-frequencies for a pentatonic scale! One can tolerate simultaneous notes from xylophones because they are extremely short transients, but if one synthesises the sounds of their harmonics, and hears them as sustained pairs of pitches, it is a very unpleasant noise.

4.4 The real questions about harmonics

And so, as I suggested at the beginning of this chapter, what actually happened in the evolution of music appears to be the inverse of the theories initiated by Helmholtz, and the questions to which we shall seek answers when we look at the operation of our hearing system are rather different to the ones the Helmholtz harmonic theory would have suggested. If pitched sounds are made up of a series of harmonics, and, as we now know, our ears separate the vibrations into harmonics before coding them, why does the sensation we obtain have one pitch, that of the fundamental? There are two versions of this question because we not only do so when the sounds do approximate to simple whole-number multiple sets of harmonics; even when sounds are made up of sets of harmonics which are not simple whole-number multiples of a fundamental, we still get the pitch of the fundamental, which, one might suggest, is simply that of the slowest rate of vibration present. We obviously could not have used harmonics in the selection of the simple ratio intervals which provided the pentatonic and heptatonic scales. How did we select the simple ratios using only the fundamentals in succession? And it seems an obvious deduction that since we liked pitches in succession in simple ratios, we would also like them simultaneously, as pairs and as chords, and we do – now. But apparently music remained monophonic for a very long time indeed. And it has remained monophonic in cultures using bells and gongs, and low-tension metal strings. The use of simultaneous pitches was slowly developed using the human voice and using organs, and the big step forward in instrumental polyphony was made by using bowed gut strings. The gut strings used in the sixteenth century had harmonics much more closely approximating to the Helmholtzian ideal series, than any of the woodwind or brass of that time.

What all this suggests is that, instead of the coincidences of harmonics of two simultaneous pitches determining the simple ratios which produce the attractive pairs of pitches, the simple ratios had been selected using only pitch-frequencies from the start, but we did not like the pitches together until those harmonics of their sounds which could correspond, did reasonably correspond. It would appear to be lack of harmonicity which was restricting the development of polyphonic music. And as instruments with better harmonicity became available, polyphonic music was developed with them. The two processes were inter-acting.

The harmonic story is one of the better example of showing a correlation between two physical phenomena: the ratios of intervals and the ratios of harmonics, and arguing circularly that each determined the other, overlooking that both of them were what hearing selected independently. The more fascinating question is that if hearing selected the pitches of the scales by using only the sensation of the pitch-frequency, the rate of the fundamental harmonic, what process does it use to select good harmonicity by trial and error? That is the really the only thing that Helmholtz inadvertently did demonstrate, and subsequent physical measurements confirm it. How instrument makers and their patrons would have described what they selected, and how Helmholtz did describe the sensation selected, was 'tone'. How does the hearing system select tone? Before we can consider that question, we need a more precise statement of how we are to use the word tone.

4.5 Tone and timbre

Musicians describe the quality of a sound they hear as its tone, and because of the great difficulty in describing any sound, they are usually reduced to using adjectives such as good, poor, harsh, warm, thin, baroque, Italian and so on. The words cover objective description, personal taste and, as we shall see, subjective belief! But tone is a real sensation, and despite the difficulty, there is often a measure of agreement amongst musicians about the quality of the sounds they hear. Tone is far more than the kind of sensation someone likes; it has played a vital part in the evolution of instruments and in the development of polyphonic music. The term timbre is also used in attempts to describe the quality of sensations, sometimes interchangeably with tone, while in some cognition studies it may apparently be used for the entire sensation of a note, other than its pitch and duration. The musicians' difficulty is understandable, but one ought not to use vague terms in objective, let alone scientific, studies, and leave it to the reader to divine what is meant. If we can find definitions for the two terms tone and timbre, then the reader will be in no doubt about what they mean, at least as used in this book. The analysis uncovers some valuable things about pitch and also some of the reasons for subjective beliefs, which are not only held by musicians.

Purely fortuitously, Helmholtz made a proposal about tone which suggests a way of defining it for our purposes. His proposal was convincing in the context of his time, and it is also enshrined in the literature. His argument was that when different kinds of instrument produce the same pitch, we can readily identify what instrument is being played, and he assumed that that was because the sounds have different tone. His experiments indicated that any harmonic present in a pitched sound was a member of the simple harmonic series. His globes gave him a very poor indication of the size of any of the harmonics he could detect. The sounds of the various instruments differed consistently in which members of the harmonic series were present. In principle, bearing in mind the crudeness of his method of investigation, what he detected was correct. For example, the sound of clarinets is made up predominantly of the odd-numbered harmonics, that is to say, the first, third, fifth and so on harmonics. The sounds of wind and brass instruments do have different distributions of the sizes of their harmonics. Low-pitched flute sound has few harmonics, while the normally bowed string produces the full complement. In the context of nineteenth-century science, the deduction Helmholtz made was very reasonable. The tones of the instruments were different, the distributions of the harmonics were different, therefore the form of the harmonics was responsible for the tone, and by that we identified them. We might be permitted an aside, that Helmholtz knew quite a lot about what the harmonics ought to be like. He was familiar with the simple theory of how clarinet tubes are supposed to resonate, he had made seminal discoveries about the vibration of the bowed string, and he was familiar with Fourier's Theorem. He was not the first or last scientist who obtained the results he expected to get. In the final chapters of his book, where he discusses the aesthetics of music, he clearly believed that harmonics were all-pervading, responsible alike for tone, the attractive consonances of the intervals and for harmony, and, reading between the lines, one detects a penchant for the more vulgar aspects of Victorian harmony too.

4.6 Identifying instrument sounds

We now know that we normally identify what instrument is being played, almost entirely by starting transients, the complex patterns of rapidly changing sounds at the beginning of notes. That surprises some people; it is contrary to popular – and Helmholtzian – belief. But that is how we identify all natural sounds, noises and words, by transient pattern changes. In general, one cannot get any orthodox instrument to produce sound without giving it a 'kick'; even the most gentle start to a sound is a transient. Pickering (1989) showed that it can take a fifth of a second or more for a violin string to settle down to steady vibration, and much depends on how the note is started (Beament, 1997). And the change from one pitch to another is a transient, however legato the music is played. Percussive sound, including that of pianos, is entirely transient. A recording of a piano or guitar played backwards is unrecognisable. If we make magnetic tape recordings of people playing normal steady-pitched notes, and then remove the starting transients of the notes, the first fifth of a second or less, it is difficult to identify what instrument was responsible for the sound. We are not used to identifying instruments by the steady-pitched sound which follows the transient, and if we splice the starting transient of one instrument onto the continuing sound of another instrument we can be confused, identify the instrument by the transient and think it has rather odd tone. If an instrumentalist plays more than half a dozen notes per second, the sound consists entirely of starting transients. We can identifying the instrument from these transients, and we can discern pitches, because there is sufficient steadily repeating vibration in the transients. But transients have no regular pattern of sizes of their harmonics, and random noise is also present.

The tone as Helmholtz implied it, is the frequency and size pattern of the harmonics when the sound has become steady, because that is the only thing he could measure, and that applies equally when using modern electronic methods as described below; if one attempts to obtain the sizes of the harmonics of the starting transients of real notes, one gets a different answer every time. (The reader who asks why I describe how big sounds are, rather than use the word 'amplitude', is referred to the beginning of Chapter 8.)

4.7 Defining tone

Leaving the starting transient on one side, consider the steady-pitched sound which follows it in a note of reasonable duration. Ostensibly we are left with sound in which the harmonics are not changing and occur in a pattern characteristic of the kind of instrument and, realistically, the only part of the sound from which we can judge tone. I will therefore adopt a modification of the Helmholtz definition and say: *tone is the sensation produced by the harmonics which are present in the steady part of an instrument's notes.* That is the meaning of the term as I shall now use it.

Using modern techniques we can obtain the harmonic content of the steady part of instrument notes easily. A note is played. A microphone feeds the sound into a computer with a suitable program and analyser. It captures a sample of a dozen successive vibrations of the steady part of the note, and a diagram showing all the harmonics – their respective frequencies and sizes – appears on the screen and can be printed. As we shall see, it is wise to ask competent instrumentalists to play the notes. Such analyses confirm that the sizes of the harmonics in instrument sounds do differ considerably, and the distributions of the large harmonics which Helmholtz discovered are reasonably correct. It is never as clear cut as Helmholtz suggested. Some produce an obvious pattern of large and small; others do not. They differ in how far they extend up the harmonic series, but within that range, the complete harmonic series is invariably present, though some of them may be very small. Small even-numbered harmonics are present in clarinet sound. Strongly bowed notes of the violin family may have the first twenty or more members of the harmonic series, but gently bowed with a mute on the bridge, only the first four or five members of the series are present.

Since this gives us the details of both the frequencies and sizes of the harmonics of the tone of any note, we should be able to reconstitute the tone. If a harmonic is a pure tone, we can use a synthesiser to generate a set of pure tones with the respective rates of vibration of the harmonics of a sample, and make each of them the size that they are in an analysed note. And our hearing should not be able to distinguish the sound from the tone sensation of the steady part of the original note.

That principle was used to imitate the sound of different instruments in many electronic keyboards. And some of the inventors suggested that people could not tell the difference between the synthesised sound and the original. I think most musicians would agree that the starting transients of some of the woodwind and brass instruments have been imitated sufficiently to identify what instrument is being mimicked; but so far as the part of the notes following the transients are concerned, I do not know a musician who cannot tell the difference immediately between synthesised sounds and those of real instruments, nor will I tempt the reader about the synthesised sound of a violin from a typical keyboard. The difference between real and synthesised tone can unfailingly be recognised by people who are concerned with the sensations of real instrument sounds. The major difference between the two is revealed by further detailed analysis of real instrument sounds. The pure tones comprising the synthesised harmonics vibrate at a constant rate; the real harmonics, including the fundamental, do not. They are all continuously varying very slightly in their rates of vibration, and also somewhat randomly and relatively independently of each other.

4.8 Timbre and discriminating pitch

We can have two sounds. In both of them the *average* size of the harmonics is the same. The continuous tiny variation in the frequencies of the harmonics in the real instrument's sound is too small to produce any detectable changes in pitch. If the tone is produced by the pattern of the harmonics – of how big each is and what ratio each bears to the fundamental, then the tone of synthesised and real sound should be the same. But listeners find the sensations of the two are different. That is why I have been using the phrase *steady pitch* and not *constant pitch* throughout, because so far as both the player's and the listener's hearing is concerned, the pitch is steady. It is an interesting property of our hearing that we didn't know we could detect the effect of the minute variations until we also heard constant-frequency sound. And a valuable way in which we can use the term timbre is to say: *timbre is the characteristic added to the tone sensation by the minute random variations of the individual harmonics.* In bowed string sound it has, somewhat poetically, been described as a halo surrounding the tone (but only the good achieve haloes).

The consequences of timbre are very significant, quite apart from its contribution to the quality factor of the sensation; it appears to have been neglected in many studies of pitch, cognition and even of hearing. It is directly concerned with how precisely we are able to perceive the pitch of real instrument sound. Consider just the fundamental. Its frequency is varying from vibration to vibration: it could be varying slightly 554 times a second if that is a musically acceptable approximate rate for C#, so we cannot perceive precisely what the rate is. It is pointless to quote the 'pitches' of a scale to three places of decimals, as some books on the physics of music do. What happens when hearing such real sound may be better appreciated through the results of one of the standard ways of testing the pitch discrimination of human hearing. Subjects listen to, say, a 'pure tone' of 440 vibrations a second, which is actually being continuously changed from, for example, 441 to 439 and back to 441, twice in every second. The change in frequency is gradually increased over a wider and wider range until the subjects say they can hear that the pitch is changing. The results are very much the same for people with normal hearing, regardless of musical ability or interest. No one can detect that such a sound, which is actually changing by two vibrations every half second, is not a pure tone, at any frequency across our entire hearing range; at 2000 vibrations a second we cannot detect such a continuous swing through four or five vibrations (see Table 4 in Appendix 6). This exonerates Helmholtz over one thing, because if you listen to a harmonic with a resonating globe which cannot have a knife-sharp resonance, and the harmonic happens to be varying continuously by a couple of vibrations – at a considerably higher rate than two per second, you cannot tell by hearing that it is not a pure tone. That has implications for how our hearing system works, aside from quite how we should define a harmonic. We do not want to get into a fruitless argument about when is a pure tone not a pure tone. This is a good example of the generalisation that a physical description of sound does not specify a sensation. We can't tell the difference between a constant pure tone and a steady pure tone. We can between a sensation with constant harmonics and a steady one with, on average, identical harmonics, because the continuous variation of the harmonics produces timbre.

So there are two factors limiting hearing's ability to judge musical pitch from instrumental sounds: the lack of constancy in the sound we hear, and the limit of our hearing's pitch discrimination. And we judge pitch in two apparently different circumstances; successive pitches and simultaneous ones. But pitch is always the relationship of two sensations, and we are limited in our ability to assess both pitches. I have avoided using the word accurate, because reality may be at variance with how accurately some people believe they can judge an interval, and taste enters into the picture too.

Then we have the problem that pitch is a single sensation, and the pitch-frequency normally corresponds with that of the fundamental. Are we judging pitch on the rate of vibration of the fundamental alone, while the harmonics provide the separately assessed tone, or are the harmonics contributing to the pitch characteristic too? We need to discover more about the role of the harmonics, amongst other things because they become very important when we return to the discussion of scales, intervals and, eventually, simple harmony in the next chapters.

4.9 Investigating tone

If the sensation of tone is the product of the harmonics, those which are present, their rates of vibration and how big each is, an electronic synthesiser can produce a set of pure tones with those characteristics. We can measure the harmonics in instrument sounds and listen to the tone independent of the timbre, of the effect of any continuous minute variations in the sound.

We have a lot of information about the harmonics in bowed violin sound, and it is a particularly illuminating example to use (Beament, 1997). Like all instruments, the violin depends upon resonance. Its body has a large number of different resonances irregularly distributed across the entire range of the sound it produces. The size of any harmonic in the sound of a note of any one pitch, depends on where that harmonic lies in relation to the resonances. If a harmonic matches the peak of a resonance of the body, it will be larger in the sound, and if it corresponds to a trough between two resonances it will be much smaller. But every violin is different because every piece of wood is different, and measurements show that each violin has a different distribution of resonances both in their frequencies and their sizes. So the sizes of the harmonics in the sound of the same pitched note played on two violins will be different. That can be proved by analysing the sounds, and checked by applying pure tone vibrations with the frequencies of the harmonics directly to the bridge and measuring each sound, thus eliminating the effect of the player.

We can produce pure tones with the frequencies and sizes of the harmonics of a note from one violin with a synthesiser. It has a single sensation of a pitch with tone. It does not sound like a violin, but that is the tone; the real violin sound also has timbre, but that only serves to obscure the clarity of the tone sensation. If we create the tone of a note from one violin, and then that of the same pitched note from another violin with a different distribution of resonances, then provided the overall balance between the sizes of the higher and lower harmonics is roughly the same, we cannot distinguish any difference between the two sounds. Similarly, take any one violin and play two notes of different pitch, say, a third apart; the harmonics of one note will lie in different places, relative to the many resonances, than the harmonics of the other note. The sizes of the harmonics will be different in each note of different pitch. But the tone remains the same over a range of several semitones.

We conclude from this and many similar experiments that our sense of tone, that is to say, the sensation of a set of constant-frequency harmonics, is a very crude one. We can examine this further with a synthesiser. Suppose we listen to a set of, say, ten or so pure tones, with the sizes of the harmonics of any violin, and with a fundamental of about Middle C. While we are listening, we decrease the size of one of the harmonics to half. We are very aware that a change is taking place, and we think the tone is changing. But if we restore the original size of that harmonic, listen for a few seconds, switch the sound off, halve its size, and immediately switch the sound on again, we cannot perceive any difference between our memory of the first tone and the second tone. Our hearing is very sensitive to any change in a harmonic while it is changing, hence the significance of timbre. If the harmonics have constant frequency, the sizes of the individual harmonics can differ considerably, but they are so amalgamated in the tone sensation that all we can perceive, rather crudely, is the overall balance between the higher and lower ones. In general, if there are many harmonics, there is a strong tone sensation; if there are very few harmonics, the tone is weak and tends towards that of a pure tone.

4.10 The harmonic contribution to pitch

We listen to a set of ten pure tones again, and then remove the fundamental harmonic completely. We still hear the same pitch. This is a most important discovery. It demonstrates directly that other harmonics are contributing to the pitch-frequency sensation as well as to the tone. It is significant in real instrument sound. There are many instruments such as the viola, cello and double bass, which have weak resonance in the lowest part of their ranges, and therefore produce small fundamental harmonics. The third harmonic of the bottom octave of the double bass is always bigger than the second and much bigger than the fundamental of the notes. But the sound has a strong sensation of the pitch-frequency. The higher harmonics in any instrument's sound contribute to the pitch sensation, and strong higher harmonics contribute an important part of it. The phenomenon provides an example of the susceptibility of hearing to suggestion. More than one distinguished acoustician has looked at a graph showing the weak resonance and consequent small fundamental of the bottom of the viola's range and said that the sound is weak, instead of listening in an unbiased way to the rich sound produced by a good player. Benade (1976) made that mistake. When acousticians believe that a graph of the resonances of an instrument tells them more about tone than their ears, one does have misgivings about what they may tell musicians and instrument makers.

A far more common experience is that the loudspeakers in small portable radios and tape players are incapable of reproducing the fundamentals of bass notes as sound, but we 'hear' the bass well enough because our hearing creates it from the other harmonics.

This demonstration that harmonics contribute to the single sensation of the pitch-frequency is a matter of great interest in relation to our hearing music. I mentioned earlier in this chapter that the ear does separate music sounds into their harmonics; our auditory system then puts them back together to produce the single sensation of pitch, or we would not get that sensation; we would get a simultaneous collection of the pitches of all the harmonics. One may suggest that this happens because the harmonics approximate to simple multiples of the rate of vibration of the fundamental and that if they diverge slightly from a perfect series they are still incorporated in the pitch sensation, but make it less precisely perceivable; they may even change the perceived pitch slightly, and they will certainly change the tone. But since the harmonics all also con-tinuously vary slightly because of the timbre inherent in all sounds produced by real instrumentalists, the pitch sensation is indeed a complex band of sound, which we could call the *pitch band*. The pitch band sets a limit on the accuracy with which we can determine intervals. If harmonics diverge from the ideal series too much, we get the 'jangle of overtones', while with xylophones, bells, steel drums and the like, we latch onto the fundamental.

4.11 Harmonicity and tone

The experiments which used a synthesiser to isolate the tone sensation from timbre, used pure tones which were exact simple multiples of the fundamental; sounds with ideal harmonicity. We could investigate what happens to tone with harmonics diverging slightly from a perfect series with a synthesiser, but experiments with 'real' sounds are usually more convincing, because musicians can relate them to their own experience.

There is a readily observed difference between the sound of bowed gut or modern gut-equivalent synthetic strings, and steel ones. A string must be elastic. When pulled to one side it is stretched, and when released it contracts back to its original shape – only it overdoes it, goes beyond the resting position, and then swings back again, and that is its resonant vibration. If the only thing producing that movement is elasticity, the vibration is ideal, and the harmonicity perfect. Gut and synthetic strings get very close to this (see Pickering, 1989). But a steel string is also stiff. Whereas nylon is very flexible, if you bend a piece of steel string and let go, it returns to its original shape. If a piece of piano string is clamped in a vice with a few centimetres projecting, and the end plucked, it vibrates briefly, producing a nasty pitched sound. That is vibration by stiffness, not elasticity, and the harmonics of a stiffness vibration are progressively sharper and sharper as one goes up the series. The clamped wire's harmonics are so far from an ideal series that it produces no sensation we would call tone. Thus when a steel string under tension is pulled to one side and let go, there are two processes making it vibrate: elasticity and stiffness. The higher we make the tension, the less it is displaced, so the less the stiffness contributes to the vibration, and the nearer it gets to vibrating by only elastic forces, the nearer it approaches good harmonicity. It never quite gets there, but that is why steel strings on bowed instruments and in pianos must be at such high tension to reduce as much as possible the sharp harmonics produced by stiffness.

If we have a solid steel string under low tension and bow it, the tone is indeed horrid. As the tension is increased, the tone becomes raw but bearable, then it has a strong cutting edge which decreases until, at very high tension and light bowing, the tone is bright and attractive. Even under high tension we can get something of the edge by bowing it strongly because that increases its vibration and therefore the contribution of stiffness. These tests demonstrate how tone varies with the divergence of harmonics from the ideal series; but because they are sensations one cannot describe the tones in any other way. The bowed sounds will be 'real' sound with timbre. Once the tension of the string has reached the 'raw' sound, the single pitch sensation becomes progressively more precisely defined as the tension is increased.

So it appears that we can distinguish two characteristics in the tone sensation. One is the balance of the size of higher to lower harmonics; the other is the harmonicity. And that is all we can deduce about the tone sensation from seeing a harmonic spectrum diagram. These experiments with a bowed steel string have a significant bearing on three phenomena: human voice tone (Chapter 5), piano tone (Section 6.13) and the preciseness of the sensation of interval ratios (Sections 6.10, 11).

4.12 The selection of instrumental tone

These discoveries about the role of harmonics in producing sensations, provide a basis for the experiments during the seventeenth, eighteenth and early nineteenth centuries, when musically sensitive instrument makers and their musician friends used their hearing to select what they deemed to be improved tone by making empirical changes in the shapes of wind and brass instruments, and that what their hearing was actually selecting was better harmonicity. It would not have been the only criterion, because better harmonicity goes hand in hand with making instruments more user-friendly, obtaining 'harmonics' which are closer to those of the heptatonic scale on brass instruments, and judging pitches more precisely. Indeed, so far as I know, the only instance where this selection process has gone into reverse is in the twentieth century, with the introduction of steel-cored strings for the violin family, with their poorer harmonicity. The only excuse for using them is economy. Gut was generally used on early plucked instruments, though the Irish harp had metal ones and according to Page (1987) they were used on some plucked instruments in the Middle Ages. He says it is difficult to decide whether some words meant brass or bronze or gold, but if strings for royalty were actually made of gold, surely the instruments must have been purely decorative? Gold is the most ductile of metals and the instruments would never have stayed in tune.

4.13 Timbre and the player

Synthesised sound doesn't have any timbre. To say that timbre is created by the player is not quite the right way of looking at the phenomenon. The stability of electronic sound provides us with the best way of thinking about the difficulty of playing any orthodox monophonic instruments; they are unstable devices. An electronic vibration generator stabilises itself; switch it on, and it produces exactly the same form of constant rate vibrations every time. Keyboards aside, orthodox instruments don't produce any sound on their own. With all three classes of instrument, the sound is produced by an interaction between player and instrument. Woodwind players can vary the pitch because the instruments do not have sharp resonances; the player determines the pitch played as precisely as the lips can achieve it. The same thing applies to how the lips are used with brass instruments. With the violin family, the player can determine pitch with a finger, but it is the evenness of expert bow speed and pressure which makes the string vibrate in a reasonably stable fashion, and it is certainly not in a constant fashion. One has only to hear the quality of the sound produced by a learner or less competent amateur playing an oboe, a violin or a french horn to realise that a musical instrument is not a physical system which 'makes the sounds'. One suspects that one or two writers on the physics of music have never played anything except an electronic keyboard and a microcomputer. It takes years of practice to discover how to produce sound from an orthodox instrument so that the pitch and loudness are steady, with appropriate starting transients.

How a player can control the tone and timbre of a wind instrument is illustrated by this story of a bassoon. It is a notoriously individual device; a player has to discover what complex cross-fingering works best for each instrument. A scientist thought to overcome this. He made a bassoon from some well-seasoned old floorboards, and operated the keys by tiny electric relays connected to a programming box so that he could finger the keys logically and the machinery would do the rest. When it was played by a professional bassoonist it sounded like a good bassoon. When the scientist played it, someone unkindly said it sounded like old floorboards. When I played double bass in the Schubert *Octet* with it, I thought it sounded more like a saxophone. The moral is that the instrument makers who developed the modern wind and brass instruments must have been very good players or obtained the services of such players, or they would not have been able to improve them entirely by hearing the sounds. Hence, when one is investigating the sound spectra of instruments, it is essential to recruit the services of skilled players.

4.14 Vibrato

The valuable property of the silicon chips in an electronic keyboard is its absolute stability; it never needs tuning. What chips of that kind have so far apparently not been able to do, is vary the frequency randomly, still less make the harmonics they generate do so; that would run the risk of having an unstable system (like a real instrument!). The simple way of attempting to disguise the constant nature of the sounds is to vary their loudness cyclically, that is to say, use excessive loudness vibrato. To many listeners, this makes the sensation even less attractive.

Our hearing appears to react quite differently to cyclical loudness variation and to frequency variation. The limited cyclical variation of pitch called pitch vibrato, which can be introduced into real instrument sound, is often thought to 'enhance the tone'; provided it is carried out at only about four or five times a second, and it extends over less than about a sixth of a semitone, the listener is usually unaware of it, and it actually makes the timbre seem richer. It is one of the factors involved in producing the sensuous sound of the bottom string of a violin and viola and of high cello. If it extends over too large an excursion or is carried out too rapidly, it becomes obtrusive. Many professional singers employ obtrusive vibrato of both kinds.

It is unfair to say that all electronic instrument inventors are unaware of the shortcomings of synthetic sounds, because the makers of more recent up-market keyboards have given up the struggle and use 'sampling'. They record samples of the sounds of real instruments produced by good players, and these are sounded in response to the keyboard. Even these only have a single kind of starting transient and a single kind of tone, and on some instruments a single sample captured at one pitch is speeded up and slowed down and used over even an octave of pitches. One can fool hearing to some extent with transposed tone; one cannot alter the speed of transients this way without offence. There is no possibility of producing the range of sounds of bowed instruments by sampling. Sampling is not a new idea. An optical sound track from real organ pipes was printed on rotating disks and used in the electric Photona organ in 1936. The deception was possible because selected organ pipes have very small starting transients.

4.15 The susceptibility of hearing

Orthodox monophonic instruments do not have tone any more than they have pitch; tone is the sensation a listener obtains when a particular player plays an instrument. Obtaining any objective assessment is made more difficult by the remarkable susceptibility of even sophisticated players and listeners to suggestion.

Because of the endless propaganda about old Italian violins, it is understandable that the public at large should believe – and players and dealers want them to believe – that the Stradivarius violin has special tone. This fairytale has become embedded even more firmly than any of Helmholtz's explanations. The popular story of how an audience can be deceived is of Kreisler convincing an audience that the cheap trade fiddle he was playing was his Stradivarius (he usually played a Guarnarius), and a similar demonstration was apparently made by Heifetz at the Metropolitan Hall in New York. How much the sound is determined by the player was first shown in 1927 in carefully designed scientific tests which have been repeated by several investigators (Beament, 1997). A skilled player plays half a dozen violins, classical and modern, to a sophisticated audience, telling them which is which, and then repeats this in a random order out of sight; no one can remember the sound of them sufficiently to identify any of them. As the experiments with a synthesiser show, our sense of tone, the amalgamated sensation of the harmonics, is very crude. The strings determine the harmonicity, and provided the strings of the test instruments all produce similar harmonicity, there is not enough information for our hearing to characterise anything.

The timbre and the transients are produced by the player, and they will be the same if the same person plays the instruments. One would have thought players would be only too pleased with this acknowledgement of their skill, but despite the number of hours a day for years on end they have had to practise to reach professional standards of sound production, some of them still want to believe that their prized instruments are responsible. The confusion is that instruments differ and players differ, and the player who thinks his instrument does everything will have found one that suits his style of playing. This only reinforces the point that a player makes six very different instruments sound alike to an audience. Kreisler may have had to work hard to make the trade fiddle behave – although by coincidence some of them are very playable instruments, whereas it is said that some Stradivarius violins are not easy to play. And although Stradivarius made about six hundred violins (more than a thousand of which are said still to exist), only some fifty of them are played. There is also a recent case of a violin suddenly losing its tone when the soloist was told that it was not a genuine product of the Italian master whose name was on the label.

But in blind tests, members of an audience can tell the sound of instruments with steel cored strings from those with gut-imitating strings with a high degree of success. This is a double test. Listeners hear three steel-strung and three gut-strung instruments and are told which is which. They then have to distinguish the two kinds when the instruments are played in random order. If they could not distinguish any consistent difference in the two kinds of tone in the first half of the experiment, they obviously would not be able to discern it in the second half.

Some of us know that a solo cello is steel-strung within twenty seconds of switching on the radio (and unless it is an outstanding performance, some of us will switch it off again). But even that characteristic sensation can be susceptible to belief. A distinguished cello teacher, who had advocated gut strings all his life, went to congratulate a soloist on her wonderful tone, and was shown a new type of tuning peg which would tune steel strings. Steel strings are almost always tuned by very obvious adjusters on the tailpiece at the other end of a cello, but having seen no adjusters, the teacher had heard gut tone.

4.16 Harmonic noise and the seventh fairytale

A synthesised set of ten pure tones with frequencies like the first ten members of a harmonic series with fundamental of about 60Hz, which is close to the pitch-frequency of cello open C, produces a sensation of roughness and noise similar to the buzz that we associate with the sound of the bowed double bass and bottom notes of a cello. The noise has been attributed to the friction of bowing, but a similar buzz sensation accompanies the tone of the contra-bassoon and the bass saxophone. Analysis shows that there is no such noise in the sounds of the bass instruments; we can create it with a set of pure tones. The buzz is generated by our ears as a result of receiving the sounds. The sensation of a similar package of ten harmonics with a pitch-frequency of C in the treble staff, is accompanied by hiss. Such a hiss is accepted as characteristic of bowed violin sound; it is present in oboe sound, but just as it is understandable to attribute the one to friction, the other may seem to be breath noise. There is, however, one difference between generating the noise in our ears by low and high notes. If we limit the synthetic sound to the first six harmonics, we get no hiss with treble staff pitches; it appears when the seventh and eighth harmonics are added. But buzz is still present from only four harmonics when the pitch-frequency is below bass staff.

There is a straightforward explanation of the cause of buzz and hiss in the mechanism of our ears. In general terms, it happens because the harmonics are too close together. Hiss can be generated by two simultaneous pure tones with the frequencies of the seventh and eighth harmonics, having a ratio of $7:8$, without any of the lower harmonics at all. The interval of a second (CD) was acceptable to primitives as consecutive sounds, but it was a long time before we used it as simultaneous notes; the fundamentals and all the harmonics of two such notes have a ratio of about $8:9$ and together produce noise, while a simultaneous semitone pair, ratio $15:16$ is not readily accepted by many today. Why a buzz sensation at low pitches should appear when a sound contains only three or four harmonics, is also explained in Chapter 10; it is a reason why closely spaced notes in bass parts always sound muddy.

The better-known supposed problem of the seventh harmonic is actually an invented problem produced by the Helmholtz theory of intervals. The pitch of a Bb in the modern scale does not coincide with the pitch of the seventh harmonic; it does not correspond with the pitch of the seventh in a simple ratio scale either (Appendix 3). Brass players normally avoid using the seventh of their instruments' series of resonances because it does more or less correspond to the pitch of a seventh harmonic; Britten instructs the horn player to produce it for a special effect in the *Serenade for Tenor, Horn and Strings*, though players don't always do so. This myth about the seventh harmonic arose because according to Helmholtz the intervals of the scale were determined by the harmonics, but his theory would not explain the interval of the seventh. So the flaw in the theory was confused by various stories about the 'dissonance' of the seventh harmonic, which led to the notion that harpsichords were supposed to be plucked one seventh of the way along the string, and piano strings struck there 'to eliminate the seventh harmonic'. It was even suggested that violins should be bowed there. The struck piano string vibrates in such a curious way that it does not eliminate the seventh harmonic wherever it is hit, and actually Helmholtz himself demonstrated that the seventh harmonic was still present in piano sound. I believe Steinway pianos hit the string one eighth of the way along (Wood, 1962). My colleague Dennis Unwin who makes harpsichords says that if you pluck iron strings at exactly one seventh length you get a nasty little squeak, which is eliminated if the quill is moved about three millimetres. One bows a violin near the bridge for strong bright tone, and near the fingerboard for soft gentle tone.

If the seventh harmonic is a simple multiple, seven times the frequency of the fundamental, it will be amalgamated in the tone, and may make a contribution to the pitch sensation, just as the fifth, or any other harmonic may. We can dismiss the fiction completely with a simple synthesiser test. Set up ten similar-sized pure tones with the ideal frequencies of the first ten harmonics of the series. Remove the seventh harmonic and it is extremely difficult to decide whether that has made any difference at all. The hiss is still there, because harmonics eight, nine and ten produce it; they are closer together still. But change the seventh harmonic so that its frequency is not seven times the fundamental, and one may hear a difference in the tone.

4.17 Conclusion

This discussion suggests that harmonics play a very important part in the sensation of tone, but because of the presence of timbre in all real instrumental sound, extreme caution must be used in applying the result of any experiment using constant-frequency pure tones and synthesis to a musical context. The discussion supports the deductions in earlier chapters that music originated with the selection of consecutive sounds by pitch: the pitch of crude instruments with very poor harmonicity, or with no harmonicity at all. It appears very doubtful whether harmonics played any part in the development of monophonic music using the heptatonic scale. And there western music appears to have stuck for a very long time. A musical revolution began sometime around a thousand years ago, which led to the use of simultaneous pitches, the development of a twelve-pitch scale, and harmony: the evidence suggests that three instruments played vital roles in this. The simple organ flue pipe, the bowed string and the human voice. The first two had reasonable harmonicity from the start. It may suggest that the further development of western music could not take place until there were instruments with adequate harmonicity. Of course, that presumes that the human voice has reasonable harmonicity. We have said little about the voice so far, and that is briefly considered next; it is considered further in the final chapter.

5   The imitating voice

Like any other musical instrument, our voice consists of a vibration generator and resonators. The vocal chords in the larynx generate the sound vibration to which the large number of interconnected and oddly shaped air-filled cavities resonate; the size and shape of some of the cavities and therefore their resonances can be varied. The harmonicity of a voice depends on the form of vibration of the vocal chords. Descriptions of the process often make the same erroneous assumption that has been used in describing the vibration generators of instruments, that the larynx produces a sawtooth wave form (Fig. 5B). That vibration has an ideal set of harmonics, and hence all voices would have perfect harmonicity. In reality, the behaviour of the larynx varies enormously from one person to another, and since no mechanical system can produce an ideal sawtooth wave anyway, you can draw your own conclusion about what a larynx can generate.

Whatever harmonics are produced by the larynx, their sizes in the vocal sound depend upon which ones excite the resonances of the various cavities. In particular, we learn to vary the resonances of the mouth cavity by changing its shape and size, and select which harmonics to emphasise in the sounds; that produces characteristic *tone* sensations which we identify as vowels. In principle, an 'eee' sound has emphasised higher harmonics from a small mouth cavity and an 'ooo' sound has emphasised lower harmonics from a large cavity. The mouth cavity is not that different in size in males and females, but the fundamentals of their larynx vibrations are very different in frequency. The human voice is therefore an extreme example of the way in which harmonics combine to produce the pitch sensation, even if the fundamental is comparatively small, but the majority of the sound energy is in the harmonics which provide the vowel sensation. With a bass voice the pitch is far below the harmonics creating 'eee'.

The voice mechanism is an example of the confusion that arises from the vague way in which the word tone is used. Using the definition of tone we have adopted, the vowel sounds demonstrate how the tone of a sound changes with the change in balance between the higher and lower harmonics. The harmonicity of the larynx vibrations determines whether the voice is 'sweet', has an 'edge' or makes a crude noise. I composed incidental music for several student productions, but the only time they ever asked me to sing as well, was for Auden and Isherwood's *The Ascent of F6* which demanded a 'blues, sung by a voice broken down by gin and cigarettes'. A voice also has timbre, because no biological system can produce constant-frequency harmonics.

The physiology of human sound production, that is to say, the instruction and co-ordination of all the parts of the body which are involved, is exceptionally complex. Discovering how to make the extensive and specific variety of speech sounds that we produce is by far the most difficult thing that most people ever have to achieve. It involves a huge number of different nerves, both those giving the instructions and those monitoring the effect of the instructions and thereby allowing control by feedback. I call this discovering rather than learning, because without undervaluing the patience and skill of speech therapists, one could not tell a child what to do even if one did know; the recipient has neither vocabulary to understand nor a means of translating instruction into physiological body actions. One makes, and goes on making speech noises to a child until by accident the child discovers how to imitate the noises. There is something in common with what is always called being 'taught' to play a wind or stringed instrument, because a teacher may offer all kinds of advice about what one has to do, but ultimately the player has to discover how to use the mouth or limbs to create the appropriate sound sensations, with encouragement and criticism.

The underlying problem is that we were not naturally selected to have a noise-making mechanism which readily performs a fraction of the processes involved in speaking a language or singing, any more than we were to hear and perceive steady pitches, or for that matter to play a musical instrument. We have to acquire the ability to control the processes by continuously monitoring the sound produced, using our hearing. In speaking and singing there is a significant amount of internally conducted vibration going directly to the ears, added to the sound from mouth via the air and into the ear in the normal way. As a result, we do not hear ourselves as others hear us, and most people are surprised when they first hear their own recorded voice. The process of speaking becomes an automatic skill, and one is convinced that in many people, the 'hearing themselves' part of the process is suppressed below conscious level. Forty years ago I constructed a tape recorder which enabled me to test this on about a hundred people. The subjects wore a pair of ear-sealing headphones, and attempted to read aloud or carry on a conversation, but they heard their recorded voice delayed by about an eighth of a second. Over half the participants were reduced either to stuttering or silence. One might conclude that the others do not listen to themselves. The technology for doing this kind of test is readily available nowadays, but I strongly advise the amateur not to try it; it was subsequently discovered that it could have a seriously disturbing effect on a small number of people.

The human voice existed before any artefact producing steady pitch was discovered, and beliefs that music originated with the voice, extend from early times to the present day. But with the exception of a few involuntary noises such as reactions to pain, we learn to produce every sound entirely imitatively. Our language learning consists of a three-way process: recognising transient sound patterns we hear, imitating transient sound patterns, and associating the sound patterns with objects or actions. Since the noises we imitate are made by our parents and teachers, there are no noises in our language that we cannot make (though foreigners always seem to have some in theirs that we can't). So also the first requirement of singing is to imitate steady-pitch sounds. But no one could have imitated steady-pitch sounds with their voice until there were steady-pitch sounds to imitate. So where did steady-pitch sound originate? With the voice? It is automatically produced by blowing across a hollow plant stem as a panpipe or end-blown flute tube. And there is hard evidence of five- and seven-pitch flutes nine thousand years ago. We can certainly go back a few thousand years from then to the discovery of a single steady-pitch sensation from a tube. But since one steady pitch does not constitute music, somewhere in that period multi-pitched music started. It is just about the same time that the first human settlements appeared, consisting of maybe two dozen individuals of all ages. No one has the least idea at what point in this period an individual sang one – or two – steady-pitched notes, and passed this information aurally to others. The previous chapter throws doubt on the possibility that harmonics could have played a significant part in the selection and development of steady-pitched music with related pitch-frequencies. The more plausible theory at this point of our discussion is that since vocalising is primarily imitative, singing steady pitches imitated artefact sounds and did not start until after the artefacts which produce them had been invented. We return to this important matter in the final chapter after we have discussed the further development of music and the hearing system, bearing in mind that our hearing system was developed with natural noises, and to process natural noises, which are transients.

There is a novel aside. If we are concerned in language with transients – pattern changes, and with the tone of vowels, regardless of the pitch of the lowest frequency present, then the pattern change of a steady-pitched sound is that its pitch does not change. If children start learning to sing steady-pitched sounds with the same learning process that they use for speech, that is to say, how the sound changes, they could reproduce the nature of steady-pitch sound successfully, without having any concern for what the pitch actually is. A child then has to discover that a second, new ability must be acquired, which doesn't happen with speech – getting the actual pitch-frequency right, which is that of the fundamental. To complicate the process further, the fundamental is not the frequencies emphasised in the tone of the vowel sounds. In a class of small children, there are usually one or two who can sing steady pitches, but not the ones appropriate to the context. They are sometimes called drones, and one school years ago used to label them 'drone' in singing classes. I hope nothing like that happens today.

This matter is also important because singing a few notes is a standard requirement of some Theory of Music examinations, which contain remarkably little actual theory of music, and should really be called the Conventions of Western Orthodox Notation. It is so easy to assume that if someone cannot sing in tune they must be 'unmusical'. Some very musical people cannot do so; Toscanini was one of a few conductors so lacking, and the inability of musicians to whistle in tune is well known.

The majority of voices do not have good harmonicity; they have tone which may be compared with that of a primitive undeveloped wind instrument. The notes have a wide pitch bandwidth. And if the almost inevitable vibrato is cultivated, it is sometimes impossible to discover what pitch is intended. The suggestion that the violin is the nearest thing to the human voice is a monstrous slight on a wonderful instrument. Voices with as good harmonicity as the bowed string, free of intrusive vibrato and possessed by singers with an impeccable sense of pitch, are almost as rare as rocking-horse droppings.

Although voices played a part in the early stages of polyphonic music, it may be argued that imitation continued to apply. Learning melodic lines aurally is imitation. Organum, which is singing in fifths, dating from perhaps AD 900 is imitation. There were attempts at crude harmony from 1100, but the earliest successful purely vocal 'polyphonic' music was rounds and canons – a single learned melodic line repeated by all parts. The problem for singers wasn't a catch but a Catch 22. One cannot obtain parts aurally by listening to music sung in parts, and parts had to be learned aurally. Notation and polyphonic music was developed with organs. Both made it possible for singers to learn parts, and correspondingly, singing in three-part harmony appeared by the end of the thirteenth century. Until the development of the bowed string, the organ was the only instrument which could produce a selected set of simultaneous sustained steady pitches, and central to the next advance in western music.

6. Hearing simultaneous pitches

6.1 The medieval situation

We can now consider how western music developed by using notes of the seven-pitch scale as simultaneous intervals, and the problems into which it ran when extending it to a twelve-pitch scale, bearing in mind that if the deductions we made in Chapter 4 about harmonics are correct, we have to seek alternative explanations for the selection of intervals.

The period between 1100 and 1500 was remarkable for musical experiments. Polyphonic music was developed; it was virtually dependent on the evolution of a viable notation system. As the reader may know (perhaps in rather greater detail than the author), notation started with signs associated with words as a reminder of where pitch should go up or down without indicating how much, then pitch changes were indicated by signs on horizontal lines, eventually standardised on five spaced lines representing the heptatonic scale of pitches CDEFGABC, the set of ratios selected by hearing on monophonic instruments. The staff system is irregular in that some steps from line to space and space to line are one semitone and some are two. But it is logical to hearing, because the scale can be positioned anywhere in relation to the staff lines by clefs. In other words, it is implicit that intervals are ratios. When further pitches were added, they were indicated by accidentals (see also Appendix 1). The duration of pitches was first shown by a heterogeneous collection of signs attached to the pitch indicators, but the chaotic state it got into was revolutionised by the invention of the bar and barline; its implications are considered in Chapter 11. Once the notation system did correspond with the way hearing perceived the pitch relationships, and with the time patterns in which composers' cortices used the pitches, the rate at which music developed from around 1500 was little short of explosive.

6.2 The organ

The organ is of great antiquity but it was a long time before it could make its special contribution. If the description by Vitorius of the hydraulus is to be believed, it had a keyboard in the first century AD with levers not dissimilar to a modern keyboard but about twice the size. It may also have had one twelve-pitch octave (Stroux, 1999, and see Section 6.3), but if so all such developments seem to have been lost before the early European development of the pneumatic organ. Significantly, in the medieval period, the organ was in places where people could read and write and develop notation, but initially the mechanism made it monophonic. Organs were built by blacksmiths and needed their proverbial strength to play them; the harmonious appellation occurred much later. Keyboards were a set of slides; one hand pulled a slide out to sound a note and pushed it back while the other hand pulled out another slide. Legato playing on an organ has always been a technical achievement. When slides were replaced by levers, the mechanism was so heavy it required a fist to depress them. The initial exuberant line of development was to increase the number of pipes sounding each note. The tenth-century organ of Winchester cathedral was reputed to have four hundred pipes, with ten pipes per note; there were two players so that simultaneous notes could have been played if there was enough wind. Contemporary accounts say that it could be heard all over the town, and people put their hands over their ears; the pop-group amplifier has a precedent in sound level and reaction to it. One can understand why the long-suffering organist who devised the rank selectors called them stops. The stop enabled musicians to hear simple flue pipes properly. The development of a keyboard operated by fingers had many implications. Because it represented pitches visually, it was important in the development of notation, but it also gave rise to the confusing terminology.

6.3 The discovery of a twelve-pitch scale

Monophonic instruments were devices to produce one octave of the heptatonic scale, and in essence they still are. The five larger intervals between successive notes of that scale were revealed when a keyboard presented a set of fifteen identical levers for two octaves of a major scale of 'C', the pitches of which could be selected by hearing. What organists really wanted was a few heptatonic scales such as F, C and G to match melodic lines with voice ranges. With two octaves of a C scale they could play FGABCDEF, and it 'doesn't sound right'. The B is not like the F in CDEFGABC. Hearing demanded an additional pitch between A and B to get a corresponding scale of F. The new pipe was called B♭ in most countries. Further experiments revealed the other gaps and the heptatonic scale was eventually expanded into twelve pitches per octave. There is evidence that the scale was extended in that piecemeal way in Europe by adding pitches. For example, a drawing dated 1511 shows a German organ keyboard with entirely white keys except for one black B♭ (though in Germany it is called B) and Praetorius wrote that such keyboards were still in use in 1620 (Blumenfeld, 1980). But given the premise that musicians' hearing wanted the sensations of the heptatonic scale intervals, it was as inevitable that experiments would produce some form of twelve-pitch scale as that they had produced the pentatonic and then heptatonic ones. If indeed a hydraulus had a twelve-pitch scale of pipes in Classical Greece, and experiments of a totally different kind outlined in Appendix 3 also led to the same scale, this reinforces the argument.

There is plenty of evidence that most organists didn't and probably couldn't exploit a twelve-pitch scale when they got one. Before a pedal manual was invented, the keyboard was extended by a further half-dozen levers called a 'short octave'. It looked as though the twelve-pitch scale was just continued downwards, and that the 'black-and-white' levers would sound B, B♭, A, G♯ and G, but they did sound something like A, G, F, D and C. Similar short-octave basses were added to early harpsichords. These pitches, either as continuous drones or a very simple bass, would only be appropriate for a melodic line in a very limited number of heptatonic scales. When a pedal manual replaced the short octave, it produced only a heptatonic C scale. In practical terms, most organists at the end of the seventeenth century didn't use more than half a dozen 'keys' (see the next section).

6.4 The tuning problem

It was when organists sought a systematic solution to obtaining a larger number of heptatonic scales that difficulties arose. Their instruments differed in two important respects from those on which the heptatonic scale had been evolved. Once selected, the pitches were fixed. And they could fix them by listening to simultaneous intervals on flue pipes with reasonably good harmonicity. The commonest form of tuning the twelve-pitch octave, thought to have been used until at least 1500, consisted of as good a cycle of fifths as could be achieved by hearing. If it could be done with physical accuracy, it would of course produce a discrepancy where the cycle comes full circle of about a quarter of a modern semitone between the two end pitches, called a comma (officially a Pythagorean comma, because two other ratios are also called commas), and that is called Pythagorean tuning. The assumption is that starting from a C or A, a succession of fifths and fourths were set by hearing. The hard evidence is that some organs had a 'split key' at the comma with two levers operating two pipes, one of which sounded A♭ and the other G♯. We call G♯ and A♭ *enharmonic* equivalents, and the suggested classical Greek meaning of enharmonic, intervals less than a semitone, seems appropriate. The number of acceptable heptatonic scales it produced was limited, but adequate for the practical purposes of the time.

Although simple flue pipes exposed the twelve-pitch riddle, evidence which in due course will help us to resolve a secret of interval sensations is provided by the harpsichord family. Its plucked thin metal string under high tension has good harmonicity and little timbre. It produces exactly the same sound every time. It is more possible to assess some intervals of the central scale of small instruments of the harpsichord family by the sensations of simultaneous sounds than with any other conventional instrument, though whether it is quite as possible as some adherents of baroque music believe is another matter.

The earliest of the contrived tunings of the twelve pitches, which are known as *temperaments*, was called meantone. In theory, it is a cycle of exact major thirds, except that there is no such thing. The series C:E:G♯:C comes to a dead end, and the derived C is nothing like an octave of the first one. The baroque recipe was to obtain fifths by hearing, and then reduce them by an iterative process until one heard supposedly exact major thirds. If it could be achieved with physical precision, it is a cycle of fifths each reduced by a quarter of a comma. And where the two ends of that cycle meet, the difference between the 'G♯ and A♭' is twice the size of a comma. According to Wood (1962), meantone made it tolerable to play in a range of about six major and three minor keys. The Pythagorean system produced sharp major thirds; meantone had flat fifths. One can't have it both ways.

The large number of temperaments subsequently devised empirically, can be described as narrowing, or occasionally widening, specified fifths of the cycle by fractions of a comma such as a sixth, quarter or half a comma, whether in attempts to increase the range of acceptable scales, or to produce the most acceptable intervals in the simpler heptatonic keys such as G, C and F. Most early keyboard music was composed in simple keys; few people wrote things in F♯ or D♭, a blessing to those (like me) with limited keyboard sight-reading ability, who (unlike me) want to play the pieces. It is consoling that according to Klop (1974), Werckmeister had responded in 1691 to a criticism of one of his temperaments, that its unacceptable major thirds only occurred in keys 'in which the ordinary organist can't play anyway'. I've never had difficulty in playing jazz in D♭, or composing in D♭; my problem is playing from piano notation in remote keys. There are some excellent accounts of baroque temperaments, such as by Barbour (1951), as long as one regards them as a source of facts, and views any opinions about their musical merit with reserve.

6.5 The just scale

A heptatonic scale was the yardstick throughout all these early experiments, and it has continued to be the set of pitches used for a huge amount of music throughout the centuries. Its simplest form is often called just-intonation, but intonation (apart from three other meanings musicians have given the term) describes 'playing in tune', and that is for the individual listener to judge. One cannot define another person's acceptable sensations, though musicians have probably been telling others what intervals should sound like ever since intervals were recognised. Even if one believed that physically determined simple ratios must by definition be 'in tune', Table 2 shows that what should be called the just-temperament scale can't consist entirely of simple ratios. The Table need not be studied in detail but it raises some important questions.

Table 2 is a version of the just-temperament scale in which I have made the majority of intervals simple ratios, and used 'D' as an indicator of the general problem. The scale cannot have simple ratios for all the fifths, fourths, minor thirds and sixths. One of each of them must be a complex ratio, and the complex ratio is about a fifth of a modern semitone away from the simple ratio. But this is apparently the seven-pitch scale everyone wanted, used still and uses. And if harpsichordists could reduce or increase fifths by a sixth of a comma by hearing, would not the departure of such intervals from exact ratios be obvious on any instrument?

	C	D	E	F	G	A	B
smtne			E:F=15:16				B:c=15:16
2nd	C:D=8:9 (9:10)	D:E=9:10 (8:9)		F:G=8:9	G:A=9:10	A:B=8:9	
mi3		D:F=27:32 (5:6)	E:G=5:6			A:c=5:6	B:d=5:6 (27:32)
ma3	C:E=4:5			F:A=4:5	G:B=4:5		
4th	C:F=3:4 (20:27)	D:G=3:4	E:A=3:4		G:c=3:4	A:d=20:27 (3:4)	B:e=3:4
4+				F:B=32:45			B:f=45:64
5th	C:G=2:3 (2:3)	D:A=27:40	E:B=2:3	F:c=2:3	G:d=2:3 (27:40)	A:e=2:3	
5+			E:c=5:8			A:f=5:8	B:g=5:8
6th	C:A=3:5 (16:27)	D:B=3:5		F:d=16:27 (3:5)	G:e=3:5		
mi7		D:c=9:16 (5:9)	E:d=5:9 (9:16)		G:f=9:16	A:g=5:9	B:a=9:16
ma7	C:B=8:15			F:e=8:15			

Table 2. Possible pitch-frequency ratios of the Just heptatonic scale of C. The table demonstrates that one cannot have simple ratios for all the intervals of either the pentatonic scale or the heptatonic scale, by using two possible ratios for C:D, 8:9 or (values in parentheses) 9:10. All the intervals containing D, that is, two of the minor thirds, fourths, fifths, sixths and minor sevenths, have alternative values. If one uses a simple value for one of them, the other has a complex value. In most cases, the difference between the simple and complex ratios is about a fifth of a semitone but the fourths and fifths differ by more than that. If one used an arbitrary value for D, lying between those given by the 8:9 and 9:10 ratios, none of the intervals containing a D would have an exact ratio. + = augmented, mi = minor; ma = major.

6.6 Selecting intervals

All the evidence indicates that from the very beginnings of music, musicians have been able to select some interval sensations by hearing consecutive notes, whatever the harmonicity of the sounds, and harmonics could not have been involved in that process at all. Somehow, they were doing this by remembering a pitch and comparing a second pitch with that memory. The pitch of the notes of primitive instruments is not well defined, because the harmonicity is poor and, together with timbre, the sensation has a wide pitch-band; a player can't make intervals more precisely than the sensations allow, and equally well, the listener cannot hear them any more precisely. But that was music: monophonic music. Simultaneous intervals were not used until monophonic players knew what to play. When they did, the sensations of intervals with poor harmonicity were unattractive – they still are.

Simultaneous intervals produce attractive sensations when the harmonicity is good. Timbre provides a pitch-band which allows a margin in making intervals, and the pitches of monophonic instruments can certainly be adjusted from note to note sufficiently to accommodate the ratio problems of a heptatonic scale. That raises a quite different problem, of how players and singers can determine them in context. Some scientists who investigate pitch believe that intervals are arbitrary and that musicians have to learn them. They do not explain how intervals might be learned 'in context'. We will return to that problem.

But if the harmonicity is good and there is little timbre so that the pitch-band is narrow, then intervals can apparently be judged very precisely by hearing; in fact, tuning organs and harpsichords to any form of temperament assumed that hearing must be able to determine at least fifths, fourths and thirds, very precisely by hearing, before adjusting them for the particular temperament. That might suggest the possibility that in those circumstances, a musician's hearing was using a 'reverse' Helmholtz effect; that then (and only then) hearing determines simultaneous intervals by matching harmonics. Initially that seems a reasonable proposition. But can that also explain how people could then set intervals to a sixth, quarter or half a comma divergent from exact intervals by their sensations? There is a modern 'physical' method by which this can be done, using the phenomenon of beats discussed in Section 6.8 below. But no one could set divergent intervals by beats until the frequencies were known, and the rate of beating depends on a reference frequency such as A=440. It may come as a surprise that even pianos were tuned by the sensation of divergent intervals and not by beats, for some good time after formulae for using beats became available. So, could anyone with practice set harpsichord intervals to divergences of precise fractions of a comma by the sensations of simultaneous intervals?

6.7 Harpsichord interval sensations

There is a consensus amongst many harpsichord specialists that the sensations of some exact-ratio simultaneous intervals are natural, agreeable or sweet, that if they are very slightly divergent they are sparkling or bright, if they diverge much more they are smarting, cutting or edgy, and if they are very divergent they are raw or brutal. In my experience, these are good descriptions for at least octaves, fifths, fourths and major thirds, and perhaps for minor thirds and sixths. They are qualitative; as an interval is made more divergent, one quality merges into another. Put the other way round, if simple-ratio simultaneous intervals could be made to diverge by different specific fractions of a comma in that fashion, those intervals would produce different sensations, and if those were irregularly distributed amongst the various intervals from a keyboard, every key could contain different proportions of sweet, bright and edgy sensations, even verging towards the brutal. How much of it can be observed during performance is another matter. We can only discuss music in terms of what are attractive sensations to the individual listener. One view might be that a lot of baroque music is so dull it needs some different sensations to spice it up; that view has often been expressed about the way the variety of organ pipe sounds relieve the tedious nature of some organ music. It also provides a basis for a belief that if J. S. Bach's clavier was in a very irregular baroque temperament, though we don't know that it was, some parts of those of the forty-eight pieces in the most remote keys might have been as edgy in sensation as they are brutal to try to play. I don't doubt some people believe that all twelve major and twelve minor keys had distinguishable characteristics; it would be difficult to devise any blind test with an irregular temperament, because those who claim to have such perceptive ability will know the key that each of the forty-eight is in. A legitimate alternative view is that well-tempered meant equal-temperament. And I'm content to concentrate on the master's superb composition, without wanting any special interval sensations to spice it up.

I have had the unusual experience of hearing the same piece of music played in rapid succession on three small harpsichords, tuned respectively in meantone, a baroque temperament and equal-temperament. The piece had to be restricted so as to be acceptable in meantone. My private reaction was to wonder whether any differences I might detect justified the fuss. I then tried a few chromatic chords on them. There certainly were differences. Barbour (1951) says that the uneven semitones in meantone can create special effects. True, but I recalled the article on temperaments in *Grove* (1927) that in baroque tuning, some of the intervals were changed by 'as much as hearing could bear'. It was more than mine could. Music is to the ear of the listener as beauty is to the eye.

6.8 Beats

My favourite definition of the phenomenon is that beats are the things you can't hear when two sounds are the same. Examining in acoustics has its rewards. The statement is correct but it needs the other half of the equation; they are the regular throbbing sensation you can hear when the rates of vibration of two pure tones are very slightly different. The only way in which one can attempt to set pipes or strings to intervals with specified divergence from physically exact ratios is by using beats. If one pure tone is vibrating at 440Hz and the other at either 438 or 442Hz, we hear two beats per second. It does not indicate which is the higher. That can only be discovered by knowingly increasing or decreasing the rate of vibration of one of them and hearing whether the rate of beating increases or decreases. Eliminating beats makes the vibration rate of two pure tones or the fundamentals of two plucked or hammered single strings exactly the same. The reason for going a very little way into the phenomenon is that it will enable us to discuss two important features of hearing music.

If we assume that harmonics are pure tones, and that the harmonicity of the sound producers is good enough, one should be able to set strings to exact ratios by eliminating beats between coinciding harmonics. If two strings are a fifth apart, that is to say, their fundamentals are in the ratio of 2:3, the third harmonic of the lower one should coincide with the second harmonic of the higher one. If the ratio is not exact, one should hear the beats between those harmonics and if one string is adjusted until there is no beating, the fundamentals should have a frequency ratio of 2 : 3. In principle one should be able to set any simple ratio in this way. In practice the beats between single harpsichord- and piano-string unisons are obvious, and it is not difficult to hear and eliminate beats between octaves, except for the very top of a piano. Fifths need practice. Beats between fourths and between major thirds are much more difficult to hear, and one cannot hear beats from the 5 : 6 ratio of the minor third. If a piano obviously needs tuning it is almost always the departures from unison between the trichords: the sets of three strings per note, which are making the nasty noises.

Beats enable one to narrow or widen fifths, fourths and major thirds according to empirical formulae to obtain whatever artificial scale one wants. It appears simple but there are several complications in setting a scale, outlined in Appendix 5. One can now buy electronic boxes which will produce a set of pure tones in any temperament one wants, and carry out the easiest of operations of eliminating beats between a pure tone and the fundamental of each string. If one's hearing cannot achieve even that, there are devices which indicate visually whether matches have been obtained. One should, however, take heart from the closing remark of Ellis (1927) in his masterly article on tuning by beats: 'but if you are half a beat out, no one will be any the wiser!' That accords with everything we have so far said about the accuracy of hearing intervals. There is practical support for that view, because piano tuners in the second half of the nineteenth century were trained to tune in equal-tempera-ment by sensation rather than beats. Imagine the misery. They spend three years listening to simultaneous piano-string notes, in order to learn the sensation of equal-temperament major thirds, fourths and fifths. Ellis checked eight pianos tuned that way by Moore & Moore's best tuners. They were often a twentieth and sometimes nearly a tenth of a semitone out. On that evidence one might question how accurately anything can be tuned by sensation.

Since one needs considerable practice to hear beats between harmonics of fifths, fourths and major thirds, under the special conditions of tuning single strings, and harpsichord tuners often use an ear trumpet to hear them at all, they cannot play any part in the assessment of simultaneous intervals under normal conditions of hearing music. One may be able to obtain the character-istic sensation of mistuned harpsichord intervals, but one cannot then attribute that to beats if one cannot hear them. One can hear beats much more easily between constant-frequency electronic sounds, and presumably an experimen-tal psychologist like Houtsma (1995) either assumes that real instrument sounds have constant frequencies, or is insensitive to the difference between real and synthetic instrument sounds; some people are.

But can we assume that a harmonic is a pure tone for tuning purposes, but also say that instruments have timbre because the harmonics are continuously varying slightly in frequency? One does not time a single beat because one cannot; where does a beat start and end? Every manual on tuning says count the number of beats in fifteen or thirty seconds; one uses an average. Beats are not absolutely regular

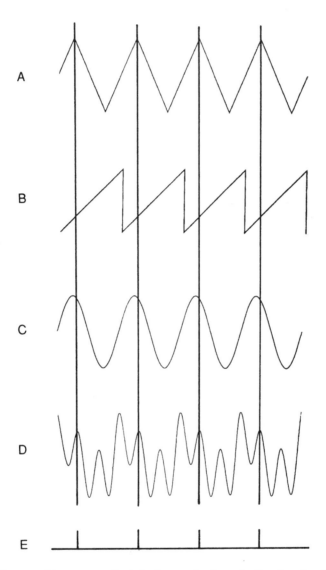

Fig. 5. The repetition rates of pitched sounds. Sound vibrating in all these different ways produce the same pitch because they repeat at the same rate. The pitch can therefore be represented by their repetition rate as in line E (see text). A: A clarinet note. B: A sawtooth (the vibration imposed on the bridge by a bowed string). C: A pure tone. D: A flute note.

but even the physical definition of frequency is based on the number of repetitions of the sound in a second and not on the duration of the individual repeated unit. We can hear beats between single piano strings, but not from a piano interval when six strings are all vibrating slightly differently. The harmonic variation in orthodox monophonic instrument sounds is so great that it prevents one from hearing beats from intervals. If one has a good cellist who concentrates on bowing evenly, one can eliminate beats between its open-string fundamental and a pure tone; it is very difficult to do this with a single stopped string, and with the slightest hint of vibrato, impossible. The most reliable way to try to measure the fundamental of an orthodox instrument's sounds is with a frequency meter, because it gives the average of ten or twenty vibrations. And one cannot get beats between two cellos playing a unison, because both strings are varying so much that there is no constant difference. If orthodox instruments did behave like typical text-book descriptions, the beating between near unisons and octaves would make ensemble music as intolerable as a piano in desperate need of tuning. If beats are as difficult to hear as they are when one is trying to tune intervals under ideal conditions, one can draw one's own conclusion as to whether they play any part in practical music. Having in Chapter 4 eliminated harmonics from most of the things attributed to them in the evolution and development of music, we can also eliminate them from judging simultaneous intervals on real instruments. They are discussed again in Section 6.11 below.

6.9 How do we judge intervals?

It is implicit in everything we have discussed so far that the pitch sensation of a musical note is a single characteristic, distinct from whatever kind of tone and timbre is attached to it. We selected simple-ratio intervals with consecutive sounds of crude instruments with poor or no harmonicity. Matching the pitch sensation of one note with another heard previously, often of very different tone, is a preferred method of tuning instruments. We can ascribe a pitch-frequency value to a pitch by comparing it in that way with the sound of a tuning fork or a siren. Regardless of the form in which a sound wave vibrates, if it does so with sufficiently regular repetition, the auditory system provides the cortex with a single sensation of pitch. In Fig. 5, four of the multitude of ways in which a sound wave could be vibrating are represented. If such sounds were analysed, their harmonics would be totally different. They would have completely different tones. One is how a strongly bowed violin string vibrates the bridge, another is the form of clarinet sound. Each is a regularly repeated form of a vibration; if they all repeat their vibration at the same rate they will all be heard to have the same pitch. Pitch is a sensation of the repetition rate. We can therefore represent the pitch sensation of the very different sound forms of Fig. 5, or of any other such sounds, by a simple marker regularly spaced at the repetition rate. We can see what is implied by the ratio of two pitch rates comprising a simultaneous interval, by spacing the markers suitably, because that is what we are actually comparing (Fig. 6). When I discuss the coding of sounds by the ear in Chapter 9 we shall find that this way of representing pitch is much closer to reality than any drawings of waves, or harmonic spectra. The shape of the wave forms represents what creates the tone. If the shape is not identical from repetition to repetition, that indicates timbre and/or lack of harmonicity.

6.10 Interval Pitch Patterns

The repetition rates of two simultaneous pitches producing a fifth are shown by the markers in Fig. 6A. It is a very simple pattern in which two repeats of the lower pitch match three repeats of the higher, the characteristic of all fifths, a 2 : 3 ratio. Figure 6B shows the pattern of the major third, a 4 : 5 ratio. Every simple interval has a different pattern, and the repetition rate of the pattern indicates the recognisable characteristic of the interval, just as the repetition rate of a single note characterises its pitch. We can use this visual representation in our discussion, provided we remember that we are using a model, and that its validity will depend on what we find when we look at the actual hearing mechanism.

Everyone obtains the sensations of the patterns of simultaneous intervals whether they recognise anything about them or not. Some people do. They do not necessarily need any musical training to do so. Obtaining the sensation of a repeating pattern does not mean that we receive a pitch which, for example, corresponds to one quarter the rate of one of the notes comprising a major third. There have been theories that intervals generate so-called sub-harmonics, but there does not appear to be real evidence that there is any *pitch* sensation which corresponds with such a pattern rate (but see Section 7.6).

When discussing temperaments, we have described hearing, for example, perfect fifths, fifths which diverge by fractions of a comma, and ones which are as wide as hearing can bear. But hearing has never been in any doubt that *they are all fifths*. That suggests that intervals, or at least certain intervals, are single characteristic sensations analogous with pitch. But if a divergent fifth does not have an exact 2 : 3 ratio, how do we reconcile that with our model of a repetitive pattern?

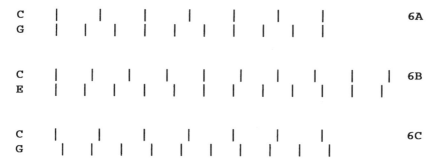

Fig. 6. The repetition patterns of intervals. A: Two pitches making a fifth [CG]. B: Two pitches making a major third [CE]. C: Two pitches making a fifth but never 'coinciding' (see text).

If we have two pure tones of 200 and 201Hz, we hear one beat per second. As a result of the form of the sounds' vibrations, the sensation we receive can be described as a loudness effect (See Chapter 10). If we start when the two vibrations will be increasing and decreasing in exactly the same way, their vibrations add together to make a large vibration. They gradually get out of step so that after half a second they are vibrating in opposite directions and cancel each other; then half a second later they come back to moving identically. They reinforce each other, cancel each other and then reinforce each other. Effectively, they are a loud sound decreasing to zero and then increasing to a loud sound again. If the two sounds come from different sources, when their respective vibrations increase and decrease are independent of each other, but there will always be a moment when the two are exactly in step and half a second later, completely out of step. Compare with that, two pure tones making a major third, with vibrations of 400 and 500Hz, and an interval pattern repeating at 100 times a second. We increase 500 to 501Hz. Like the mistuned unison, the interval pattern will be 'out' halfway through each second, and 'in' at the end of each second. But we do not hear any beats.

In Fig. 6A the model indicates the repetition pattern of two pitches comprising a perfect fifth, with two of the markers coinciding, indicating that at that moment both vibrations have reached a maximum. The interval pattern repeats at 100 times a second. But as Fig. 6C indicates, if the 200 and 300Hz pure tones come from different sources, their maxima need never coincide; yet the interval pattern is repeating 100 times a second, whether maxima ever coincide or not (in this sense, hearing is not phase sensitive). It is still a perfect fifth. And if our hearing says it is the same sensation regardless of whether any one vibration of the 300Hz coincides with one of the 200Hz vibrations or not, how do we know that the pattern relationship is changing from matching to not matching once per second, when the higher pure tone is 301Hz? The answer is that we can't tell by using repetition patterns because over any fraction of a second, there is a repeating pattern sufficiently corresponding to a fifth. This is quite different to the production of beats. And it is therefore impossible to judge the exactness of intervals produced by two pure tones; aural tests using them are not valid.

When we use real notes with harmonics we have a separate source of information; it was most clearly demonstrated by simultaneous notes from a small harpsichord. As an interval diverges from an exact simple ratio, it is the *tone* which changes. We can only describe tone qualitatively. The individual string has sweet tone, and two strings in an exact-ratio major third also have sweet tone. It can't be sweeter than the tone of the individual strings. As the interval becomes wider, the tone becomes 'bright'; if this was a change of a quarter of a comma – a sixteenth of a semitone – it could not be detected with two pure tones (Appendix 4). As the string interval is further widened, the tone becomes edgy, then smarting, cutting and eventually brutal. But it is still a major third! If we now refer back to the tone of a bowed stiff steel string under different tensions (Section 4.11), when the harmonics of an individual string are progressively very sharp, their tone is raw and brutal. As the tension is increased and the harmonics become less and less sharp, the tone changes in a very similar fashion, through cutting, edgy and eventually to bright. Under the highest practicable tension and very small vibration, it might even be described as sweet. It is said that the way to determine the ideal tension for a harpsichord string is to raise it by a semitone and sneeze; if the string breaks it was at ideal tension.

The inference is that one obtains a particular tone sensation from one set of harmonics which diverge from a simple whole-number series, and a similar one from two sets of harmonics which individually have excellent harmonicity, according to the amount they diverge from the exact ratio of some intervals. Both produce the tone of divergent harmonics. And if two notes each have poor harmonicity: edgy or raw, then the interval tone is going to be edgy and raw or worse even if the fundamentals happen to be in an exact simple ratio, and it gets worse still if the ratio diverges.

But if we identify the sensation of two simultaneous sounds as a 'fifth' regardless of whether it is a flat or a sharp fifth, that is to say, divergent by a considerable margin from a reasonably exact 2:3 ratio, then 'fifthness' is a sensation in its own right. Simple tests with musically experienced people (though not experienced in tuning steel strings to intervals), suggest that the range over which hearing refers simultaneous pairs of notes to the nearest interval differs between intervals. The octave, fifth, fourth and major third interval sensations extend virtually from a quarter-tone flat to a quarter-tone sharp, and there appears to be a point between the major third and fourth where the sensation switches from an excruciatingly sharp third to a very flat fourth. There is some uncertainty about the boundary between the minor and major third, and between the augmented fifth and the sixth, and greater difficulty about where the sensations of semitones, seconds, augmented fourths and sevenths begin and end.

6.11 The reality of simultaneous intervals

The whole of this discussion of music has been based on the initial premise of Chapter 3, that some people, a minority of any community, select the sensations, and that it is for the majority to decide whether they like those sensations. I have been discussing that minority throughout. Certainly, to us, the sensations of octaves, fifths, fourths, major and minor thirds, the augmented fifth and the sixth, are real phenomena of the hearing system. We can characterise them even when they diverge substantially from simple ratios, and also when the harmonicity is poor though we don't like the tone sensation. We can do so when the sounds have substantial timbre and modest vibrato. This cannot be based on matching harmonics, or on beats between harmonics. And although no doubt some people recognise these things rapidly nowadays from the experience of hearing them in an accessible form, one certainly cannot agree with Houtsma (1995) who apparently believes that intervals are an arbitrary convention which have to be learned by solfeggio. That fails in basic logic, because they would have to be learned from somewhere, and we reached the conclusion in Chapters 3 and 4, that we selected these interval relationships from consecutive pitches out of a random set with poor or no harmonicity at all, many thousands of years ago, and selected them by hearing on organs a thousand years ago, without knowing what we were selecting.

It would appear that the only basis on which all this can be modelled is the repeating pitch pattern. People are able to remember the pitch pattern of one pitch and compare it with that of a note heard previously, at least for certain intervals. This is why we normally tune instruments by hearing the 'A' and then matching it. We don't want the conflict of the tone of simultaneous sounds, particularly from different orchestral instruments. We often tune the violin family by consecutive fifths (fourths on the bass) though we may then check them simultaneously. And this is why many adults who are not musicians find it difficult to help a six-year-old to tune a violin 'even with a piano', which certainly doesn't produce the simplest of sounds.

There can be no denying that music based on the sensations of the octave, fifth, fourth, major third, and minor third have always been, and continue to be, by far the most widely accepted musical phenomena by the passive recipient, but it is especially valuable to have the experimental evidence of Krumhansl (1990) who showed that presumed musically naive subjects referred to the 'nearest interval', pairs of pitches which did not have exact simple-ratio fundamental frequencies. In other words, the phenomenon is passively real as well as actively real. And if the recipient obtains the interval sensations regardless of the exactness of the ratios of the fundamentals of the notes, it accounts for the relative insensitivity of a great many people to what musicians call intonation. Many of the notes produced in pop music preclude such refined judgement, but they still convey the simple intervals adequately for adherents. But even the most pitch-sensitive musician can only determine intervals to the extent that tone, timbre and the pitch accuracy of hearing will allow. For that reason I never use the so-called unit of pitch, the *cent*, which is a unit of frequency and not of pitch. It gives a completely false impression of how accurately we can hear any pitch. I am sure it has caused anxiety amongst musicians because it suggests that they ought to be able to hear pitches far better than they can. We cannot hear even a pure tone to better than 10 cents over most of the range of pitch-frequencies used in music. A detailed explanation of why the cent should be abandoned is given in Appendix 6.

6.12 Equal-temperament

The way out from the problems of varying sized intervals is the compromise of making all the intervals of any one kind have the same ratio, and hence equal-temperament, which was suggested as early as 1630 by Mersenne, and perhaps by others. Yet it only began to be adopted for general use in the nineteenth century. It always seems to be described as making all the semitones equal, which is probably the least thing in which musical hearing is interested, and it is certainly no help to most musicians to be told that the ratio of the equal-tempered semitone is the twelfth root of two. Equal-temperament distributes the comma evenly between all the fifths of a complete cycle so that each is one twelfth of a comma less than a physically exact fifth. A twelfth of slightly less than a quarter of the tempered semitone is almost exactly a fiftieth of one, a mighty small difference; likewise, the major thirds are one sixth of a comma larger. The ability of hearing to discriminate pure tone pitches is given in Appendix 4 together with a comparison of the simple and equal-tempered ratios.

Under suitable listening conditions in isolation, hearing can distinguish the divergence of an equal-tempered major third from a simple-ratio one, from a small harpsichord. It is doubtful if it could from monophonic instruments, assuming that either ratio could be exactly achieved by players, or that it could distinguish the two versions of the fifth or fourth. The other intervals raise the same problem that we saw in Table 2. Have they got an exact value? If they can have different ratios in the just-intonation scale, with what does one compare the equal-tempered ones?

Whether listeners can, or believe they can, distinguish keys in baroque music played on small harpsichords tuned to an appropriate temperament, keys do not differ in equal-temperament. Baroque beliefs persist. In one month on BBC Radio 3, presenters introduced piano works in the 'sombre' key of C minor, the 'noble' key of C minor and the 'romantic' key of C minor. Whether the music merited any of those adjectives, it had nothing to do with the key. It is sometimes difficult to persuade otherwise rational people about keys. One argument which is often effective is to point out that A might have been almost a semitone lower or higher than 440Hz when pieces were written. Equal-temperament is, of course, a totally artificial scale, and it is unacceptable to use it as the norm against which to assess the divergence of just-intervals.

It is often helpful to explain the 'simple' principle of the equal-tempered scale visually. The spacing of guitar frets shows it. Another way is by a drawing of a set of panpipes (Fig. 2B), because the lengths correspond with the pitch-frequencies. After one of my friends expressed surprise that the ends of the pipes were not in a straight line, he asked if it was any special kind of curve. 'Of course,' I replied, 'it's a logarithmic curve.' His face lit. 'And you mean to say I've been playing in logarithms all these years without knowing it?' Molière's bourgeois gentilhomme would have embraced him. Sadly, I had to disillusion him. He plays with superb intonation and no drawing can show that.

6.13 The piano

The piano illustrates many points in the previous sections. Within a period of about forty years ending around 1800, the piano ousted the harpsichord as the fixed-pitch workhorse of music. True, it had properties, particularly in the control of dynamic, which the harpsichord lacked (Cole, 1998). It could also produce louder sounds. Because of Cole's unbounded enthusiasm, I think he does get things a little bit back to front. It was not new styles in composition which advanced the position of the piano, but the properties of piano sound which made new types of composition acceptable. As for the piano having a sweet tone, most advocates of baroque sound might disagree. A major property of the piano is that the complex sound it produces prevents anyone from being able to judge the exactness of its intervals. When a steel string under very high tension is struck by a hammer, a curiously shaped kink runs along its length and back again in a cycle at the rate of the fundamental. That generates a vibration in which there is random variation in the harmonics. And when three identically tuned strings are struck together, they interact with each other, primarily through their common bridge, producing wide timbre. Hence no one can distinguish its equal-tempered ratios from simple ones from the sensations. It is the fixed-pitch keyboard instrument which combines acceptable harmonicity, timbre, a wide pitch-band and equal-temperament.

There is a parallel between the sensation of the piano trichord and that of a bowed string section. I cannot be sure whether the aphorism that 'one violin's a violin, two violins is out of tune, and three violins is a string section' is my invention or whether I heard it years ago, but many string players agree. Two good violinists in 'unison' don't produce sufficiently matching pitches such that one will not hear an occasional difference, but when there are three or four, their combined timbre and slight pitch differences make it impossible to distinguish each; like piano sound, the temperament is immaterial.

The piano does have a small range of tone change related to how hard the strings are struck. The louder sound does have stronger higher harmonics, but also if a stiff steel string is displaced very little, the harmonics are less sharp and the sensation is sweeter. When the strings are firmly struck, the harmonics are sharper and the tone is brighter. But that can be exacerbated by the form of tuning. After a central octave of pitches has been set (see Appendix 5), the upper strings are tuned in octaves. Theoretically an octave is a simple ratio to determine, but many people like slightly sharp octaves. And slightly sharp octaves are 'bright'. The higher octaves of a piano are invariably tuned very slightly sharp. Our piano tuner loves playing our Bösendorfer after he's tuned it, and says, 'Listen! Isn't it bright.' I haven't the heart to tell him it's because he's tuned the intervals bright. In contrast, when first hearing the sampled piano notes of my Roland electronic keyboard, which is in exact equal-temperament, many musicians say that the higher notes are flat. Have some people been acclimatised to the brightness of real piano intervals from childhood, and adopted that as the pitch ratio sensation? I asked the violinist Tessa Robins, who plays with superb intonation, about pitching octaves, and she said she could hear them better if she played them very slightly sharp.

Since all descriptions of the tone sensation are qualitative, and depend on an individual's taste, a piano lover can describe its tone as sweet, bright or anything else. But it was the form of piano sound that allowed a keyboard instrument to indulge in the harmonic variety of the romantic and later eras. It also had a special role in making music of any kind more widely available in domestic circumstances until the advent of the gramophone and then radio.

6.14 Consecutive intervals

It is clear that a soloist on any monophonic instrument must use the mechanism of remembering the pitch pattern of notes in order to determine the pitches of other notes. But if we look again at Table 2, do solo players playing in C major play two different seconds, and how do they deal with an augmented fourth or the sevenths? There are at least two possible answers to the question. One is that if the interval is indefinable, such as the augmented fourth, neither player nor listener can object to the intonation, and this can be applied to other intervals such as the major seventh. The other is that in orthodox tonal music, the mysterious meaning of tonality is that musicians do not simply apply pitch-pattern memory sequentially, pitch by pitch through a melody; what happens when people do is illustrated in the following Section. Tonality means that we carry a base-line memory of the pitch pattern of significant notes of the diatonic scale (hence its name), that we are aware of all the simple-ratio pitch patterns of Table 2, and we always know where notes and intervals lie in relation to the current 'key'. We do not need to have 'absolute pitch' to do this; illiterate jazz musicians do it automatically. We do not have to learn solfeggio; it's only people who can't do it automatically who have to learn such things to sing.

It would follow from this that at least so far as the simple-ratio intervals are concerned, any new synthetic scale which has intervals which diverge from those ratios, will be regarded by musical hearing as out of tune. It does regard intervals which are half a semitone different, often called quarter-tones, as excruciatingly out of tune. It seems likely that hearing does endeavour to refer such intervals to simple ones as with, for example, the aberrant pitches of the bagpipe scale (Section 3.9). And I'm afraid that Wellin (1991), who wrote that quarter tones are the next logical step in the series, was not only wrong on physiological grounds but failing both in musical knowledge and simple logic. The sequence pentatonic, heptatonic, dodecaphonic is not a series. If our ancestors had selected the augmented scale CEG♯C, developed it into a whole-tone scale CDEF♯G♯B♭C and then into the dodecaphonic one, which is a series, the next logical step would be quarter-tones, but it would not be the next musical one.

6.15 Intonation and pitch stability

Helmholtz (1870) claimed to have demonstrated that a singer produced different pitches according to the interval sung, though no one today would accept his experimental method. It is a nice idea that when an experienced string quartet plays a chord of C major it is closer to simple intervals than equal-temperament, and if that is followed by a chord of, for example, E major that is likewise closer to simple intervals. The problem about any such belief is implicit in the cycle of fifths: that if each pitch is determined only by reference to the previous pitch, the overall pitch will almost inevitably drift. Vocal groups do. In 1949 I accompanied a motet on a lute in a play in the open air at Queens' College. The producer decided at the last minute that the extensive middle section be unaccompanied. Anyone familiar with the identically tuned guitar will appreciate the feeling of panic when your singers leave you in G major and return in F♯! The pitch stability of some professional vocal groups is impressive. Yet it is rare indeed to hear the unaccompanied vocal quartet in Verdi's *Requiem* sung with acceptable intonation. Most soloists are a law onto themselves.

So how does a symphony orchestra of eighty or more people, all playing instruments on which they each individually determine the pitch of each note, maintain stable pitch? In one sense, they behave in an analogous fashion to a biological system; if an interacting mass is big enough it is self-stabilising. The answer to all of these ensemble problems in orthodox polyphonic music is that one plays by a combination of using one's own tonality and listening to the result of other people's tonality. Even very experienced players can run into difficulties when sight-reading complex tonal music; my chamber music produces the occasional cry of 'I can't hear what I'm meant to play.' It isn't only advanced tonal music which can produce such problems. Some of the Purcell *Fantasias* are amongst the most difficult music to play, because it is modal (for which I have never found an adequate definition either). And if the music does not provide tonality? We have to become automatons.

7.   Patterns in harmony

7.1 The basis of harmony

We can now briefly discuss the phenomena of more complex relationships of the interval sensation which are broadly covered by the term 'harmony'; at the end of the chapter, a few other discoveries about harmonics are mentioned briefly. They do not have any direct role in music, but they raise questions about our hearing system which have a bearing on the origin of music.

Experiments using simultaneous pitches in polyphonic music revealed a remarkable aural phenomenon, the sensation of three simultaneous pitches. If adjacent pitches are rejected, the heptatonic scale produces six triads: [CEG], [GBD], [FAC], [ACE], [EGB] and [DFA]. The sensations of these triads, as simultaneous pitches: *chords*, are apparently amongst the most widely acceptable and enjoyed sounds man has discovered, mostly by people who have no idea of what they are hearing.

In the previous chapter I have argued that musicians remember pitch-patterns and compare them with those of successive pitches. It is consistent with that view that the three pitches of any one of these triads heard in succession can establish the provenance of the triad as effectively as if the three pitches are heard together. Monophonic music in sequences of the pitches of one or other of these triads abounds, and it is not important in what order the pitches occur. To many musicians, the sensations of pitches of one triad in any sequence identifies that triad just as the chord itself does.

However, [CEG], [GBD] and [FAC] are given the distinctive title of major triads, while [ACE], [EGB] and [DFA] are called minor triads. Why do we put the first three in a class called major and the other three in a different class? The text-book answer is that a major triad consists of a major third [CE], plus a minor third [EG], whereas a minor triad consists of a minor third [EG] plus a major third [GB]. Then why do they sound different? Few musicians produce a succinct answer to this question and often one does not get any. The terminology is typical visually based rubbish, which is further exposed by what are called inversions, for if we rearrange the order of the pitches of a major triad as [EGC], that is now a minor third [EG] plus a fourth [GC]. But it is still a major chord. To generalise, three pitches such as C, E and G, can be in any order and it is still a major chord. There is only one answer to the question: because one can hear it's a major chord. And similarly if the three notes A, C and E are distributed in any order and I hear

Fig. 7. The repetition patterns of chords. A: A major chord [CEG] ratios 4:5:6. If the pitch-frequency of C is 400Hz, this pattern repeats 100 times a second. B: A minor chord [ACE] ratios 10:12:15. If the pitch-frequency of C is 400Hz, this pattern repeats 33.3 times a second.

it, I will say that's a minor chord – because it sounds like one. But what sensation is being identified?

All the reader need do is appreciate the principle of the following paragraphs; there are further observations in Appendix 7. 'Major chord' is a characteristic of a particular kind of sensation which can be obtained by hearing it, just as can the simple intervals. There is no reason why anyone should identify a major chord sensation when listening to music. But if it is recognised, what is being recognised is a sensation produced by three pitch-patterns which are in the ratio of 4:5:6. The major third ratio C:E is 4:5, the minor third ratio E:G is 5:6. The suggestion that one adds intervals to make chords, propagates the same misconception one finds in some music dictionaries, that intervals are differences in pitch. The vital factor in such a triad is that as well as a 4:5 pattern sensation and a 5:6 sensation, a 4:6 sensation C:G is produced. And since intervals are characterised by their ratios, a 4:6 sensation is the same as a 2:3 sensation – a fifth. It is yet another case where the hearing system uses the logic in the sounds, but the terminology obscures it.

The easiest way of looking at what happens with chords is by using the distribution of markers representing the repetition patterns of the component instrumental sounds, as we did earlier with intervals. The major triad [CEG] with ratios of 4:5:6 produces a repeating pattern (Fig. 7A). If the three pitch patterns are, say, 400, 500 and 600 per second, they have a common repeating pattern at 100 times a second. If we have its inversion [GCE] as 300:400:500, the common pattern repeats 100 times a second and it does so in the inversion [EGC] as 500:600:800 too. All three forms of the major chord have the same rate of repeating pattern. It will also, separately, have a tone which depends both on the tones of the component notes and on how closely the pitch-frequencies approximate to simple ratios. We can identify a sensation as a major chord if the three pitches have different tones, though we may not always find it an attractive sound even if their pitch-frequencies are close to ideal ratios.

We can apply the same principle to the minor chord sensation [ACE]; it is a minor third [AC] 5:6 and a major third [CE] 4:5, but [AE] has the 2:3 ratio of a fifth. The simplest form in which we can write the ratio of the minor chord is

$10:12:15$; $10:12 = 5:6$, $12:15 = 4:5$ and $10:15$ is the $2:3$ ratio of the fifth. This (Fig. 7B) has a much longer repeating pattern in relation to the pitch-frequencies of the component notes. [ACE] is called the relative minor of [CEG]. The pattern of the relative minor repeats at one third the rate of the relative major. The pattern characteristic makes it possible to identify any particular harmonic sensation, using the ratios of vibration rates of three (or four) pitches to each other (Appendix 7).

The principle of the creation of repetitive patterns from combinations of pitches works just as well if instead of having the pitches close together, one of the pitches is a bass note and the other two members of a triad are a couple of octaves higher. [CEG] as $100:500:600$ still has a repeating pattern of 100 times a second. The Helmholtz harmonic theory provided a explanation, that 500Hz and 600 Hz correspond with the fifth and sixth harmonics of C=100Hz. But try applying his harmonic theory with E two octaves below C and G. The third harmonic of E is B, its fifth harmonic is G♯; neither produce any useful relationship with the C or G two octaves higher.

The pattern-repetition rate of a major chord provides a logical basis for the relationship of simple chords in succession. The repetition rate of the chord [CEG] with ratios of $400:500:600$ is 100 times a second; the repetition rate of the dominant [GBD] with ratios of $600:750:900$ is 150 times a second. The relationship of the chord patterns, $100:150$ is $2:3$, the relationship of the interval of a fifth. The pattern rate of the sub-dominant [FAC] is related to [CEG] as $3:4$, the interval of the fourth. In other words, the repeating pattern rate of the commonest chords are related by just as simple ratios as those of the intervals.

When polyphonic music first developed, how restricting was it to use only the sensations of the six triads obtainable from the seven pitches of the heptatonic scale in sequences or simultaneously, and confine the harmony of the music to those three major and three minor triads? The remarkable thing is that even when the restriction was removed by the development of practicable baroque temperaments, and the many pitch combinations of the twelve-pitch scale were available, how few composers made use of them. The baroque period has a handful of outstanding and inventive composers, but an astonishing number of minor ones (most of whom seem to be mentioned in the *New Oxford Companion to Music*), whose compositions are 'little known'. There is usually one very good reason. Of course, a lot of that music was not written to be listened to, but to be played whilst people ate and conversed. We know that Telemann's *Tafelmusik* was, but he was such a gifted composer compared with most of his contemporaries that some of it is worth hearing.

Bar after bar of an enormous amount of baroque, classical, popular and pop music, as well as traditional monophonic music for dancing handed down through the aural route, consists of pitches relating to one or other of these six triads and especially the alternating use of [CEG] with [GBD]; that, commonly called 'tonic and dominant' harmony, has been the basis of naive music ever since polyphonic music became established.

Orthodox polyphonic music was developed by an extension of this process of relating it through sequences of triads, and other sets of pitches which were selected from the twelve-pitch scale, whether heard as the sensations of chords or not. It requires a minimum of three pitches to provide a uniquely identifiable chordal characteristic, such as major, minor, augmented fifth [CEG♯] and so on. The number of different combinations of three, four and even five pitches which can be selected from an equal-tempered scale, and which produce unique patterns which are identifiable harmonic sensations generally acceptable in orthodox music, is actually rather small. There is little evidence that the majority of formally trained instrumentalists overtly identify chordal sensations as they listen to music, any more than they consciously name intervals as they play their instruments, though in ensemble playing, as discussed in the previous chapter, it is normally an inherent awareness of the relationship of the pitch-patterns they receive, with those they play, which maintains pitch stability.

The possible application of the relationship of pitch-pattern rates to more exotic chords and chord sequences is discussed briefly in Appendix 7. But it is not the aim of this investigation of music sensations to attempt to develop a theory of harmony alternative to that produced by Helmholtz (1870), which has flaws in it. The aim is to establish that there is a logical basis for the sensations of intervals and simple traditional harmony and, in Chapter 9, to see if there is a corresponding basis in the mechanism of the hearing system. The more complex the relationships of the pitch-frequencies, the more complex the information presented to the cortex in the sensations, and the less simple the pitch-pattern relationships will be. If anyone wishes to examine all the relationships of advanced chromatic tonal harmony in detail I suggest they start by compiling a table similar to Table 2 but showing all the ratio relationships between the pitches of an equal-tempered twelve-pitch scale extended over two octaves, in pairs, threes and fours. It would need a large piece of paper. It would be a waste of time, because it all happens automatically in the cortex of those who compose tonal music there, and to various extents in the cortex of those who play tonal music, whether they have had any formal musical education or none.

7.2 The harmonic sensation

It follows from the normal way in which we use the hearing sense, of using short-term memory in order to obtain information from sound sensations, that we can recognise not only the matching of pitch patterns in sequences and associate them with harmonic sensations, but also extend this to recognising it from a melodic sequence which contains other pitches, provided the minimum of three pitches of a 'chord' are present. We can use the patterns of pitch-frequencies regardless of the other components of the sound. If the pattern is strongly represented in the fundamental of notes, as for example from steel drums, it can be used to produce readily recognised chord sensations, despite the absence of any harmonics which are simple multiples of the fundamentals.

One may not like the resulting tone, but some naturally very musical people do. One might suggest that in popular music, tunes are presented with an accompaniment of chords, because that provides the recognition of the harmony for the recipient.

7.3 Harmonic shorthands

Sequences of harmonic patterns are just as memorable and recognisable as are tunes. There have been two systems of representing them in harmonic *shorthands*. They reveal a remarkable parallel between performing baroque and jazz music, but they serves to emphasise the common basis of all tonal music.

In the baroque period a keyboard player was provided with the notation of a melodic line, usually an extremely sparse bass line, and a figured bass, a set of numbers which indicated what chordal sensation to create until the next set of numbers occurred. It had great advantage so far as the composer was concerned. It was a very rapid way of writing down compositions in musical shorthand, and it enabled a skilled player to interpret pieces with a florid accompaniment if an Italian audience liked it, or a more austere one if that was suited to a German audience, for the same piece. It treats a harpsichord player as a special kind of musician, presented with a harmonic framework and expected to improvise on it just as does a jazz musician. Figured bass is ideal for music using the harmony restricted by baroque temperaments. It can be used for chromatic harmony, though it becomes more difficult to sight-read. It cannot, however, deal with enharmonic changes (where a piece changes 'key'), since it always relates to the basic heptatonic scale of the piece. It is to be doubted whether the advance of chromatic and enharmonic harmony spelled the demise of the figured bass. The French must accept some responsibility, because each printed piece was prefaced by instructions even on precisely how every ornament was to be played, which is the very antithesis of what attracts a musician to improvisation. Only 'harmonic' musicians were able to play direct from figured bass and they did not grow on every tree; there are many such musicians today, but few of them are attracted to baroque music. Many baroque pieces have now been realised, that is to say, someone has written in a collection of dots so that the unable can play a version of the pieces, though it denies the entire spirit of the enterprise in the same way that trying to play jazz from dots does. If a composer writes a piece in detail, it makes the player a slave of notation, but one cannot have a magnificent piece of Bach's contrapuntal writing without it, even if many of the harmonic progressions he used are stereotyped to the point of predictability.

Harmonic shorthand returned with the chord symbol. I have been unable to find mention of this invaluable system in any proper academic reference book though it is included without explanation over the examples of tablature in Read (1969), but any beginner's tutor for guitar or banjo has it. It was in use in the 1920s when a little four-stringed guitar called the ukulele enabled people to play chords, the harmonic basis of the popular music, a role overtaken by the guitar. The chord symbol is comprehensive. It specifies any chordal sensation in a maximum of five letters and figures. As with figured bass, it does not specify any notes; that is left to the player. If needed it can be used as a basis for jazz improvisation. I discovered harmony with a ukulele when I was seven years old; I found I could listen to any popular music on the radio and then play its harmonic sequences, invaluable when I migrated to banjo and guitar. I didn't discover how to read notation for another dozen years. The habit of 'reading' the harmony as one listens to any music is, like absolute pitch, a mixed blessing, for it leads to automatic reading the harmony of all the music one hears. And one is constantly assailed by banal music in almost every public place these days.

7.4 Learning harmony

Since harmony is a sensation, the only way to acquire a facility is aurally. The problem is that there is no more point in playing one chord to a child and saying 'that is major' than to point to a wheel and say 'that is round'. If you present a child with a wheel, the full moon, a dinner plate and a coin and say 'they are all round', you still have to leave it to the child to observe the sensation of roundness, and similarly if you play a succession of major triads, the recipient has to divine the 'major' sensation. It puts across entirely the wrong message to point to dots or letters and say 'that is a major chord'; you don't draw a red cone with green speckles and tell anyone it tastes of strawberry, and you can't tell anyone what strawberry taste is.

I have no idea of how a knowledge of harmony is obtained otherwise. I have discussed it with many professional players, and the academic subject appears to be rather like acoustics, something to be dismissed from their minds as soon as they have passed what, if any, examinations they have had to take at college. If it is based on the half-dozen typical textbooks I have, I am not surprised. It appears to be part of the general question, can music as distinct from technique be taught, or is it something one acquires?

7.5 Other simultaneous-pitch phenomena

There are other phenomena arising from the sensations of simultaneous pitches which we may include in our questions for the hearing system. The first of these is the *Difference Tone*. It has been used by organ builders since at least the seventeenth century. If two pipes with fundamentals making a fifth, for example of 150 and 100 vibrations per second, are sounded together, one also hears a pitch-frequency of $150 - 100 = 50$ vibrations a second, an octave below the lower pipe. It is therefore an inexpensive way of obtaining a deep bass note and saving a great deal of space. A possible explanation can be based on the experiments described in Chapter 4, which showed that a set of harmonics could produce the sensation of a fundamental when there was no fundamental in the sound. In the example given above, one pipe produces harmonics of 100, 200, 300, 400Hz and so on. The other pipe produces 150, 300, 450Hz and so on.

If the two sets of harmonics acted together, the series would be 100, 150, 200, 300, 400, 450Hz and so on. That could be regarded as a harmonic series of a pipe of 50Hz with its first, fifth and seventh harmonics missing. It could generate a missing fundamental.

The organ builder's difference tone is produced by two pipes a fifth apart. Two other circumstances produce pitch sensations when there is no sound of corresponding frequency present. They are produced by loud pure tones. The more interesting one produces a sensation which has a pitch the difference between the frequencies of the two pure tones, regardless of their frequencies. It is heard even if one pure tone is sounded in one ear and another pure tone in the other ear. It therefore must be generated where the signals in the nerves from the two ears come together, which they do several times on their way to the cortex (Fig. 25B in Appendix 8). This is also called a difference tone, and it avoids confusion if we call it the *neural*-difference tone. Since the neural-difference tone between 150Hz and 100Hz will also be 50Hz, there has understandably been argument about what causes the 50Hz sensation. From my own experiments, it is doubtful if one can hear a neural-difference tone if the higher of the two pure tones is above about 2.2kHz, which may fit in with a significant feature of the hearing system (Chapter 10).

The third phenomenon is that two very loud pure tones heard in one ear produce a number of very faint other tones, with pitches related to the frequencies of the two tones, of which the most easily heard does not correspond to the simple difference in frequencies. In all probability these are caused by overloading the mechanical system of the ear and are a product of distortion. It is extremely unlikely that they play any part in the phenomena of music, despite suggestions in earlier literature.

It is possible that one or both of the genuine difference tones are responsible for an odd phenomenon recounted by del Mar (1983) that when woodwind players produce certain chords, the string section of an orchestra may ask who is playing the other note. The one thing of which we may be reasonably sure is that in none of these phenomena is anyone hearing a harmonic, though that is how it is often described.

Chording is requiring all those notes which should start at the same time, to do so; it is one of the features of a fine orchestra. Orchestral string players are used to doing this, because sections play in unison and so 'chords' are similarly expected to be started with precision. The term is normally applied to obtaining precision when three or four wind or brass instruments have to play a chord. The technical difficulty is that each instrument has to be kicked into speaking, and there is a big difference between kicking a clarinet and kicking a bass trombone. A remarkable feature of our hearing system – though not everyone exercises the option when hearing music – is that if we hear a simple major chord played by three horns and any one of the three produces its starting transient sufficiently ahead of the other two for it to be identified, we go on hearing that instrument. This is called *tracking*, and it is also important in following, for example, what are always called the inner parts of a quartet. But if any three instruments start sounding at exactly the same time, the only way in which we can tell that there are three instruments playing is the chordal sensation. One cannot get the sensation unless there are.

7.6 The three-tones paradox

Putting a new sound sensation into short-term memory is vital to the way we use hearing in obtaining information about any kind of sound. It is a cortical process, and therefore presumably we learn to do it. A phenomenon I call the *Three-Tones Paradox* raises interesting questions about harmonics. In the experiments describing the synthesis of tone with sets of pure tones in Chapter 4, the component sounds were switched on together. If we generate three pure tones, for example of 200, 600 and 1000Hz, we can regard these either as the first, third, and fifth harmonics of a note with pitch-frequency 200Hz, or as three pure tone pitches related as C, G an octave above, and E in the next octave: a spread major triad. If these three tones are switched on together, we obtain the sensation of a pitch-frequency of 200Hz with a tonal characteristic produced by the harmonics. If they are switched on in sequence with a second or two between them, we obtain the sensation of a major triad. This is an the extreme example of tracking; if our short-term memory system obtains a sound it continues to track it as a distinct entity. If it is received with other harmonics, it is amalgamated. Like many experiments with pure tones, it must be done with headphones, because with this special sound, our hearing may switch from one to the other form if there is a sudden change in loudness of one of the harmonics because of room reflections, described in the next chapter.

It may be compared with another experiment which demonstrates tracking. This consists of hearing a synthesised set of pure tones with ideal harmonicity, and while doing so, a pure tone with the frequency of one of the harmonics is sounded. It is heard as a distinct entity. When that pure tone is switched off, the subject continues to hear 'that harmonic' as a separate identifiable sensation. Tracking it when it is added is not so surprising. It will have a starting transient and be a 'new' sound. Continuing to track that frequency when the additional source is switched off suggests that once identified, the brain has distinguished a specific sound and it is still there. How the hearing system appears to works with simultaneous pitched sounds is discussed in Chapters 9 and 10.

7.7 Can we hear harmonics?

Harmonics were the basis of previous theories about the origin of western music. They are crucial to our understanding of tone, pitch discrimination and other matters. They are also involved in, for example, the belief that music originated with the human voice, which is discussed in the final chapter. Synthetic sounds compounded from pure tones representing harmonics of ideal and non-ideal series have been used in a very large number of very carefully designed experiments on hearing, and indicate what subjects can or cannot distinguish *in such sounds*. But we have to be extremely cautious in extrapolating the findings to hearing music under normal listening conditions; apart from other things, orthodox instruments and voices do not produce constant frequency sound. We are concerned with what normal listeners hear. What people believe they can hear after they have been told that they can is a very vexed subject (Section 12.5). Harmonics cause corresponding vibrations in the ear's mechanism, and the ear sends coded information about them into the brain. But as we shall see in Chapters 9 and 10, the evidence appears to support the view that from normal musical instrument sounds the brain does not, with one exception, receive harmonics as independently signalled *separate entities*. Under special circumstances we may hear them if, for example, the harmonics of bells or cymbals decay at different rates, but otherwise they are amalgamated in the two sensations of pitch and tone. This is discussed in detail in the chapters on the hearing mechanism.

8. Loudness

8.1 The basic dynamic scale

Detecting how big a sound is, is the most primitive ability of our detecting system. If the reader wonders why I have not used the term *amplitude* before, it is for the same reason that we cannot talk about the frequency of a pitched note. It has several frequencies and we now know that some of them contribute to the pitch sensation too. The term for the excursion through which a pure tone vibrates is its amplitude, but since all noises and musical sounds consist of several frequencies, each with a different amplitude, it is best for the time being to stick to words with ordinary meanings such as big and small. All the frequencies in a sound contribute to the sensation of loudness. And our ancestors probably detected roughly how big a noise was before they had any sense of loudness; some animals with no cortex can. Loudness itself is a sensation.

We have no absolute assessment of loudness; like almost every other modality we can sense, our judgement is comparative, whether a sound is louder or less loud than the previous one. The rough scale which our hearing system automatically provides is that the sound vibrations have to increase three times to double the loudness sensation, at all parts of the loudness range. Starting with low level sound, to be twice as loud the vibrations must increase about three times, ten times to be four times as loud, thirty times to be eight times as loud, and so on. Why our hearing behaves that way was very valuable. We needed to know whether a noise was getting louder, probably meaning getting nearer, and an increase in the vibrations of a tiny noise that we needed to detect, would be a negligible increase in a big one. It means that an instrument has to increase its sound output by about as much to go from *f* to *ff* as to cover the range from *pp* to *f*, something it is difficult to get young orchestral players to appreciate. But it does not mean, as even one or two otherwise respected acoustics books suggest, that ten violins in unison will be four times as loud as one. The rule applies to increasing a single sound, and several independent sound sources do not add loudness as steeply as that. Hence one needs a very large chorus to compete with a full orchestra.

In playing music we refer to the broad level of loudness as the dynamic, the actual sound levels depending on what is producing the sounds. Using a dynamic range is a comparatively recent innovation. Simple wind instruments are not capable of much range; they are unstable devices. There is a wind supply level below which they do not sound at all, and then a level at which pitch changes with wind pressure; who has not heard the excruciating moan of an expiring bagpipe? At too high pressure, many wind instruments become unstable again. The recorder, which is really still a baroque instrument, has a very limited loudness range. The inability of the baroque bow to produce high pressure, and the lower string tensions of viols, limited their dynamic range. The harpsichord mechanism produces a single level and kind of sound. One of the less recognised achievements of the eighteenth and nineteenth century instrument developers was to devise wind and brass instruments which players could make stable over a wider dynamic range, though it still is not very large. The modern bowed string has the largest range. Music for dancing has always been at one level, and with the advent of the electronic amplifier, often an ear-damaging level.

8.2 Loudness and frequency

The thing which complicates our judgement of loudness is that it differs greatly according to frequency. It is true that when one asks subjects to judge when two pure tones of different frequency are equally loud, and when one pure tone is twice as loud as another of the same frequency, the agreement of those with normal hearing is surprisingly close. Our hearing is often said to accommodate a phenomenal range of about one to ten million in the amplitude of a pure tone, but that only applies to our most sensitive range, around 2–3kHz , and it is due to our extreme sensitivity to low level sound in that region. Our range is much smaller with low and very high frequency sounds, because if the vibration is small, we cannot hear them at all. Once we can hear them, their loudness increases steeply with increasing amplitude, so that at what is called the threshold of pain, similar sized vibrations are equally loud across the whole hearing range (see Table 9 in Appendix 16). No single harmonic even begins to approach that level in music played on orthodox instruments of any kind.

The comparative loudnesses of pure tones of different frequencies has been used in the best sound level meters to provide some indication of loudness, but they can only provide either an average or a peak value. The way a sensitive meter on a good tape recorder flickers about when recording music gives an inkling of how the sound level varies, and the detailed calculation of the apparent loudness of a single musical chord is very complex. The loudness of each component harmonic contributes, but large amplitude harmonics make a disproportionately large contribution, and that also depends on where they occur in our hearing range. At normal levels of instrument sounds, the higher harmonics up to about 4kHz primarily determine the loudness sensation, and the higher pitched instruments are, generally speaking, louder. In physical values, a double bass probably produces twenty times as much sound as a violin, but they have similar loudness.

8.3 Loudness and tone

The simple way in which one can experience our sensitivity to different frequencies over the hearing range is to listen to a recording of orchestral music at a realistic level, and then turn down the volume control. That reduces the size of all the vibrations by the same proportion, but it progressively reduces the loudness of the lower frequencies compared with those in the 1–4kHz region. The balance changes and the sensation has less bass. It can be restored by using the bass tone control which on any well-designed amplifier increases the size of the lower frequencies to compensate for the properties of our hearing, increasing the lowest frequencies most. On the other hand, while changing the balance by electronic means affects the attractiveness of the general sensation of orchestral music (and that depends on an individual's taste), a change in absolute sound level makes surprisingly little difference to the realism of an individual instrument, provided the notes are above about middle C. A particular example of this is the solo violin in a concerto. The sound radiated from the instrument is diluted in the volume of the auditorium. The actual vibration of the sound a listener receives from a violin 30 metres from it in a concert hall is less than a thousandth of that the soloist hears, but provided the instrument has strong resonances which produce strong harmonics in the most loudness-sensitive range of our hearing, it will be well heard, and if well played will sound like a violin. This phenomenon is commonly called 'projection' and the same term is applied to singing. Sound cannot be projected; it travels at a fixed speed and its audibility to the recipient depends entirely on how much energy the source puts in the frequency range where our hearing is most sensitive. Of course, apart from being much louder, the balance of the harmonics will appear quite different to the violinist, with an ear only a few inches from the radiating surface. The higher harmonics will not be as dominant at such a high level. Hence one cannot discover whether an instrument will project by playing it. Only distant listeners can find that out.

Several matters arise from this. High violin, flute, clarinet, piccolo can be heard above the full orchestra. Their low notes, with just as large vibration, are easily swamped, the bass clarinet especially so. The skilled orchestrator takes this into account. Very high notes may have their fundamental and second harmonic where hearing is very loudness sensitive, but they don't have and can't have much distinguishing tone because the majority of the harmonics are above the useful part of our hearing range, and they tax the player because hearing is especially sensitive to both pitch and the regularity of the sound at such high altitudes.

8.4 Loudness and pitch

Although high sound levels appear to affect the pitch of some pure tones, with any complex sound such as music, the pitch sensation is independent of loudness, otherwise the volume control of an amplifier would not work. This is an exceptionally important feature of our hearing so far as music is concerned; at this stage of our discussion we cannot advance any reason why it should be so, and it will be particularly interesting to see whether our hearing mechanism explains how the two characteristics are separately sensed, and, if possible, why they are.

8.5 Varying loudness

It is characteristic of our senses that we detect changes while they are happening much more readily than we can make assessments of static situations, and our sensitivity to fluctuating loudness is in extreme contrast to the rather rough levels of dynamic we use. A particular example of this is the way we detect what are very small loudness variations in what is otherwise a 'steady' note. One element in what is colloquially called poor tone is irregularly fluctuating loudness due to lack of playing technique. We are also very sensitive to loudness vibrato. And the experiments with a synthesiser (Section 4.9) demonstrated how easily we detect the change in amplitude of one of a set of harmonics. Measurements under laboratory conditions show that we can distinguish very small differences in loudness between steady pure tones. People with normal hearing can distinguish more than 300 levels of loudness of a pure tone an octave above treble staff. How we achieve such discrimination is discussed in Chapter 10, though the even more interesting question is why we have such sensitivity to small changes, and yet generally use only very rough contrasts in dynamic in playing.

9. Music through the hearing machine

Our hearing system selected artefact sounds with particular characteristics that appear to have been the basis of pitched music from the beginning. What we would really like to know is whether our hearing system provides reasons for liking such sounds, but no one can explain that, despite the vast literature which purports to do so. All we can ask is whether the hearing machinery presents sounds of the selected kinds to the cortex in a form which reveals its unique characteristics, and would facilitate selection. Does it, for example, show the patterns of simple intervals and chords we have described?

This account of hearing is a small selected part of a huge, complex and sometimes puzzling body of knowledge, in which there have been major discoveries in the second half of the twentieth century and some within its last decade. In his book *The Psychology of Hearing*, Brian Moore (1982), whom I hold in great respect, ends one chapter with a reference to two other books and adds 'but you may find their interpretations are somewhat different'. What follows is how I interpret our knowledge of the hearing system specifically in relation to music. It is certainly over-simplified, and it may suffer the same fate as previous attempts to relate scientific discovery with music put forward by Helmholtz and by others, but even if it does, it will not change anything about music I have discussed in the other chapters. In science we continuously change our ideas when new facts are discovered. Perhaps one day, musicians will change some of their beliefs too.

I think the way in which music is described by orthodox terminology and in textbooks is more difficult to understand than anything in this chapter and the next one. But as a result of the discussion so far, I hope the reader may already have a different if not a better understanding of music, regardless of whether what follows provides satisfying answers to the questions which have been raised in the previous chapters. One obvious feature of music: rhythm, is missing. I have left that until after an account of how we hear, because we shall then at least understand why hearing is so sensitive to time.

9.1 The questions

How do our ears detect sound, and in what form is the information about it sent through our brains?

Is steady-pitched sound peculiar to the hearing process?

Does the process distinguish harmonics? If it does, why don't we hear them?

Does the process distinguish the lowest rate of vibration, or the lowest common factor amongst harmonics, as the pitch, and if so, how? What happens if sound has unrelated harmonics?

Can the system explain why tone is an ill-defined sensation?

Does the mechanism show a relationship of steady pitches in simple ratios?

Is there a basis for interval and simple chord patterns? Why are intervals of a second [CD] less attractive simultaneously, and semitone intervals [CC♯] unacceptable simultaneously? What is the cause of bass instrument buzz and high instrument hiss?

Why should intervals be wide in the bass but can be narrow in the treble? Why is our sensitivity to sound and our pitch accuracy so different in different parts of our hearing range?

Why are the pitch-frequencies we normally use in music in the 60–1000Hz range when we hear best in the 2000–4000Hz range?

Why are we so sensitive to transients?

Does the mechanism distinguish steady-pitched sound with timbre from constant frequency sound?

How can we track individual instruments in polyphonic music?

What is the basis of loudness?

What is the basis of 'stereo' and 'surround sound'?

Why should our hearing system have such features, without which we would not have music?

9.2 The evolutionary background

Our hearing is a system which produces very detailed sensations in our cortex when sound enters the ears. That is how we use it, but the only reason we can do so is because we have an advanced cortex. Since that is how we always experience sounds it is difficult to appreciate that the sensations provide a version of the physical sound in a particular form and not an exact replica of it. Music depends entirely on the form in which the sensations are delivered. It is even more difficult to appreciate that we are using automatic machinery, the result of millions of years of evolution, which was not in fact designed to provide detailed sensations. Other higher mammals have very similar automatic machinery (de Reuck and Knight, 1968) and could get similar sensations, but the extent to which they can use them depends on their cortices. When the proud dog-owner boasts that 'he understands every word I say', some dogs may recognise the sensations of two dozen words, and a chimpanzee can do better. In this and the following chapter there is a description of the machinery and how it appears to work, especially in relation to music sounds, but the key to understanding hearing is to solve the problem of why we and our animal relatives have this automatic system, if it was not for producing detailed sensations. That provides the real answer to all the questions.

Hearing is the result of what our very distant ancestors needed for survival in a natural world. A vibration in the environment could mean danger, and a hearing system started as a simple device which sent a message to a primitive brain to put the body on alert when the device was vibrated. It still does; an unexpected sound in silence causes alertness and the release of adrenalin. The first ear they developed also had the great value that it detected sound coming from any direction, and when the source of vibration could not be seen, and our ear still does.

The next valuable development was a system which assessed the size of the vibration. If the sound increased, the danger was approaching. A special unit was developed in the primitive brain to assess the messages in that way. It needed an ear that sent messages in a form which could indicate the size of the vibration. The unit was the first specialised part of the brain, developed to produce automatic reactions to a feature of sound. It is of course the origin of our sensation of loudness. If you are interested, such a unit is described in Appendix 9, though I suggest you read it after Section 9.5.

Finally, this hearing system was developed so as to indicate the direction from which a sound came. It was enormously valuable for survival because it produced the automatic reaction of which way to scarper! It also pointed the head or the eyes or rotating external ears in the direction of a sound source. We can observe that happening with many animals. And we are provided with the directions of sounds automatically. We may think we don't need such survival mechanisms today, but people who listen to loud music while walking, cycling or driving do lower their chances of survival.

Making a direction-finder with our type of ear was very difficult. When a motorist asks how to get to a destination in central Cambridge, one is often tempted to say 'well, I wouldn't have started from here', and that was the problem with this hearing system. It started with a type of ear which does pick up sound from every direction; that was the first priority. It needed a very complicated set of automatic units in the brain to determine sound direction from them, and that required the ears to signal the sound in particular ways and in great detail, so that the units could divine direction and produce appropriate reactions. That appears to have been the main factor determining how our ears now detect sound, the form in which it is signalled, and, therefore, the form in which our cortex gets the sensations (Fig. 8).

We get the sensation of transient sound in detail because the direction-finder needed the detail. That enabled us to develop language millions of years later. It needed the time of arrival of sound to be sensed very precisely; that is why time can be used significantly in music. The system operates on transient sound because natural sound was transient. When we accidentally discovered how to make steadily vibrating sound a few thousand years ago, the hearing system processed it and it fortuitously emerged in the form in which we all get its sensations; the cortex received and developed music sounds in that form. That is why music exists. There are other kinds of ear which provide sound direction very simply, but do not provide any details of

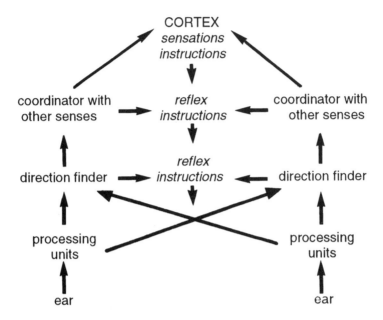

Fig. 8. General form of the hearing system. Signals from each ear pass along its auditory nerve to automatic processing units in the brain. Those from both sides are compared in direction-finders, which can send instructions to the eyes and other organs. The signals are co-ordinated with the input from other senses and can also cause reflex reactions. They then enter the cortex where sensations representing the sound occur. The cortex can both send instructions and over-ride the reflex reactions. See Figs 25A and B in Appendix 8 for greater detail.

the sound. If our ancestors had started with that type of ear, developing language and music would have been almost impossible. We would have to communicate by vision and scents. Other animals do. There are more details in Appendix 8.

9.3 An overview of the hearing system

In the ear, vibrations caused by sound distort minute processes on special living cells, and this produces streams of electrical pulses in the tiny nerves connected to them. The pulses signal information about the sound. Initially we can think of nerves as minute wires; they can be a metre long or less than a millimetre. They convey pulses from the sense organs to the brain, and instructions from the brain to parts of our bodies to do things. The brain works entirely by and with such signals. The signals are rather like a Morse code, except that they are all dashes – there is only one kind of pulse. Everything works by the pattern of the streams of pulses, where they come from and where they are sent. Pulses initiated at one end of a nerve, travel along it, and cause corresponding pulses in one or several other nerves with which they make contact, forming our communication network, the nervous system. Transmitting information as streams of pulses in digital recording and computers is commonplace now, but you don't need to know anything about how such things work to understand this chapter.

The individual tiny nerve fibres from the vibration-sensitive cells are gathered into a bundle called the *auditory nerve*, and the signals in the individual nerves pass through the series of anatomically identifiable automatic processing units, which together we call the *auditory pathway*. The signals cause the automatic reactions described in the previous section as they pass through, and then go to the cortex (see also Figs. 25A and B in Appendix 8) where they produce the sensations. We are able to suppress some reactions. If someone hits a drum behind us, we jump; if we see someone hit a drum, we don't.

We know quite a lot about how ears work, and a great deal about nerves, because all nerves appear to work in very similar ways. We know in very general terms, the functions of the processing units. But when the pulses pass into the cortex, all we know is that they create sensations. We are reduced to mapping which regions are active when different senses are used, and the parts which are active when someone hears noises, or speech or music, but even those regions are not the same from one person to the next. The cortex is so complex that the most we may ever hope for is to understand it in principle, since the evidence we already have suggests that no two cortices work in precisely the same way. That is called individuality; people have a lot of it.

Our other main source of knowledge comes from 'black box' experiments; the term arose from using boxes of electronics without knowing what happens inside, and that's most of us. Black box tests regard everything from the outside of the ear to the cortex as an unknown box of tricks, and use what sensations we obtain when we hear special artificial sounds. They are invaluable when carefully designed and tell us what our hearing system can and cannot hear or distinguish, especially when hearing pure tones and similar simple sounds. The reason for using pure tones has nothing to do with music; if we use the simplest of sounds, and ask very simple questions, such as 'can you hear that' or 'is X louder than Y', the answers tell us properties of the hearing system. It is clear from previous chapters that music sounds are not pure tones. If we want to investigate music we have to do it with real musical sounds. The experiments have shown that if people have undamaged hearing, they appear to receive similar sensations from the same sounds. Our hearing range, how small a difference in pitch between two pure tones we can distinguish in different parts of that range, our judgement of loudness and so on, are apparently very similar in everyone. The science of psycho-acoustics and a lot of medical audiometry would be in difficulty were it not so. This differs from vision, which often needs correction as soon as defects can be assessed, and from taste, for some people experience no taste from substances which are very bitter to others. So far as we know, there are no such things as golden ears or musicians ears. If a composer claims to hear 'differently', he must not expect anyone else to do so. Sadly, an enormous number of people have damaged their hearing by listening to very loud music; a few have faults in the wiring, an example of which is given in Section 9.27.

9.4 The hearing range

It is usually said that our hearing starts at about 16Hz, an octave below the pitch-frequency of double bass low C, but that is not strictly true; we can detect a sound vibration down to 4Hz or lower, but below about 16Hz a pure tone produces a sensation of discontinuous blips. If you have a means of generating a pure tone below 16Hz, never listen via a powerful amplifier. Use headphones at low volume, and remove them at once if it produces any strange sensation. In some people it may interact with the alpha rhythm of the brain. The continuous sensation starts at about 16Hz, and we can deduce an important fact about the ear from this (Appendix 15). Our hearing is very poor at low frequencies; at double bass low C pitch-frequency, a pure tone vibration has to be more than 2000 times as large as one a couple of octaves above treble staff in order that we can hear it at all, and when the bottom octave of a bass is played pianissimo, we would not hear the pitch unless it was created by the other harmonics. Loudness, in the sense of how comparatively loud a pure tone is for a given size of vibration, increases, together with pitch discrimination and loudness discrimination, continuously up to an octave or so above treble staff, and these remain excellent between 2 and 4kHz. Then all these abilities decrease quite rapidly as we approach our upper limit of hearing: 16kHz or more when we are young, gradually dropping to 9kHz or less in old age, though there is considerable variation.

The different parts of our hearing range have different roles, well illustrated by the telephone. Speech was intelligible when telephones only transmitted sound up to 1kHz, but it was difficult to recognise who was speaking. We can identify most individual voices on the modern telephone which transmits sound up to 4kHz; it would not add significantly to recognition to hear frequencies above that. But frequencies down to at least 300Hz are essential to make speech intelligible, and our hearing can provide lower pitches from voice sounds as it does with the double bass, and with small portable radios. Very high frequencies are important for transients and percussive sounds. They contribute much less to pitch. The sensation is sensitive to the balance of frequencies and therefore recording technicians have considerable responsibility which they do not always exercise to listeners' satisfaction.

So if our hearing is most sensitive from about an octave above treble staff upwards, and if we were to believe that music is concerned with observing sound sensations in detail, why are the pitch-frequencies of most notes used in music well below that range, and go right down to 31Hz, where hearing is very poor? The reason is simply a practical one; save electronic methods, all the ways we have discovered for making pitched sounds would run into difficulty trying to produce pitches more than two octaves above treble staff. Solo violins and piccolos only get half way up the sensitive range while reeds and brass haven't a hope. One bit of physics is immutable; an instrument's length must be halved for every octave one raises its pitch range. The extension in range from the primitive discoveries of how to produce steady-pitched sounds have been downwards rather than upwards.

9.5 The sound receiver

The external part of our ear is a funnel collecting sound (Fig. 9). It passes along a tube into a small cavity which is broadly resonant over a range of about 2–4kHz, boosting any sounds of those frequencies. They can be detected at lower levels than any other sounds. Harmonics in that region enhance the strength of the pitch sensation in which they are amalgamated. Instruments or voices with good resonances in that range therefore project in an auditorium. The ear tube ends at the eardrum, a delicate membrane across the tube, which is vibrated by the pressure changes of the sound. Against the inside of the eardrum leans the first of three tiny bones suspended by muscles in the air-filled middle-ear cavity. The bones transmit the vibration across the cavity, and fulfil a number of functions, amongst which they help insulate the incoming sound from body noises. The third bone in the chain, called the stapes, is a piston, pumping the vibration into the liquid-filled cochlea, a tube containing the vibration detector. The vibration is transmitted along the bones with remarkable faithfulness but if the sound is very loud, the muscles suspending the tiny bones can contract, and reduce the level of vibration reaching the cochlea. There are further details in Appendix 10.

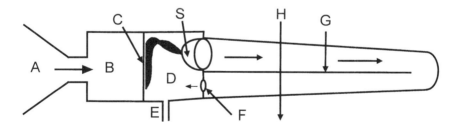

Fig. 9. The ear (schematic). The external ear funnel (A) passes sound to a resonant tube (B) ending at the eardrum (C). The air-filled middle-ear (D) contains three tiny bones, of which the stapes (S) pumps vibration into the liquid-filled cochlear tube, drawn uncoiled, containing the basilar membrane (G) which vibrates. Residual vibration passes back into the middle-ear cavity by the oval window (F). The Eustachian tube (E) connected to the throat, maintains equal air pressure either side the eardrum. Fig. 10 shows a section through the tube at any point such as at H.

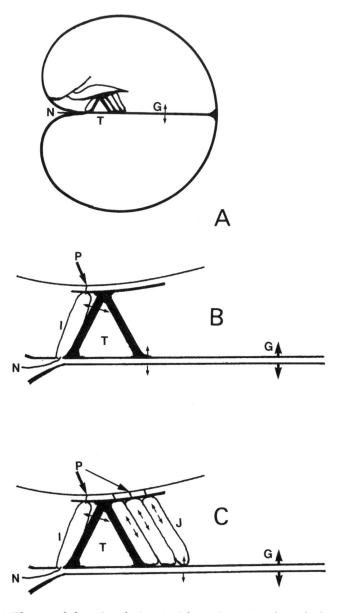

Fig. 10. The sound-detecting device. A: Schematic section through the cochlear tube. The vibrating basilar membrane (G) rocks the triangular frame (T) and the cells either side of it. B: Tilting the frame distorts the process (P) on the hair cell (I) causing it to fire a pulse in its nerve (N). C: It is believed that when the processes on the triplet of cells (J) are distorted they expand and contract, and increase the rocking of the frame. This allows the signalling hair cell to respond to much more minute vibration. See text and Appendix 11.

Fig. 11. Distribution of frequency on the basilar membrane. Values (in Hz) suggest positions of maximum resonance of the membrane to different sound frequencies (see Fig. 12). S is the stapes end where sound enters the cochlea. Maxima are roughly spaced in octaves over the middle range, but appear to be closer together at low frequencies. As the text explains, the positions are not a critical part of signalling frequency.

A cochlea is a spiral hole in solid bone, which is one of many reasons why it has been so difficult to investigate it; to explain the mechanism it is easier to think about it as a straight tube (Fig. 9). There is a partition dividing the tube into two and about 35mm long, running almost the length of the tube, called the *basilar membrane*. Sound travelling through the liquid vibrates this membrane. Along the membrane there are about 7500 *hair cells* evenly spaced in a single row (Fig. 10A and B). Each cell is attached to a tiny triangular frame. The 'hairs' are minute processes on the end of the cell, which are very sensitive to movement. They project through the top of the frame. If the frame and cell moves and this is detected by the hairs, that causes the cell to fire an electrical pulse into its nerve fibre. So the principle of signalling sound is that when vibration of the membrane rocks the triangular frames, the hairs on the cells are repeatedly displaced by it, and that produces streams of corresponding pulses in their individual nerve fibres which are transmitted to the auditory pathway.

The first question is how does the minute vibration of sound in the liquid make the membrane vibrate? The membrane is resonant; it vibrates resonantly when sound traverses the liquid. At all the sound frequencies we can hear? Things can resonate over a small range of frequencies, but no object can resonate at any frequency from 16Hz to 16kHz, or so we might think. The way this problem was solved is a major factor in our ability to distinguish sounds of different frequencies in music. A stiff membrane resonates at high frequencies; a floppy one does so at low frequencies. That is how one tunes a kettledrum. The basilar membrane is stiff at the end where vibration enters the tube, and resonates at high frequencies there. It continuously decreases in stiffness to the far end, where it is most floppy and resonates at very low frequencies. Some piece of the membrane resonates at any frequency in our hearing range (Fig. 11).

A pure tone causes its particular length of the basilar membrane to resonate and vibrate the hair cells along that piece. A different pure tone causes a different region to rock the cells there. In this way, a sound such as a musical note has its component harmonics spread along the membrane as different vibrating regions, according to their rates of vibration, the lower the frequency

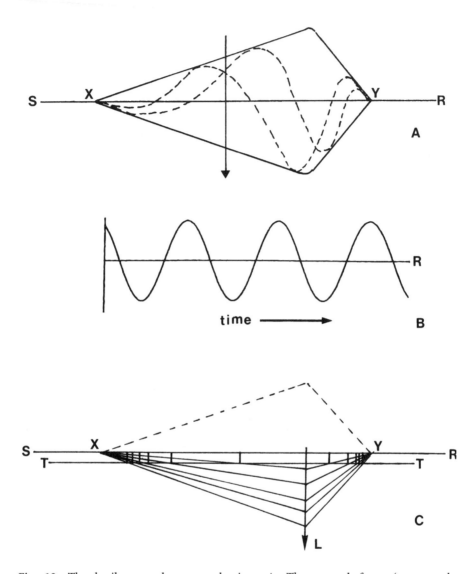

Fig. 12. The basilar membrane mechanism. A: The general form (enormously exaggerated) according to Bekesy (1960) of the wave travelling along the region of the basilar membrane which resonates when it receives a pure tone. The wave starts where the membrane begins to resonate (X), increases to a maximum and then decreases rapidly to zero (Y). Two instantaneous samples of the wave are shown within the envelope which represents all positions of the wave. The length vibrated spans several hundred signalling hair cells at low and medium frequencies, but many fewer at high frequencies (see Fig. 20). In this model the 'down' movement from the resting position R causes hair cells to fire as the wave runs past them. B. Any point along the resonating region, such as that indicated by the arrow in Fig. 12A, will have an up-and-down movement in time, in the form of the pure tone causing the resonance. C: In principle, a

of the harmonic, the further it is from the end where the sound enters the tube. These will then vibrate different groups of the hair cells at those places along the membrane.

9.6 The travelling wave

The way the membrane resonates was discovered by Bekesy (1960), after years of persistence and great ingenuity. He showed that the vibration produced by a pure tone is a ripple travelling along the membrane, which is shown enormously exaggerated in Fig. 12A. It is a little like the surface of a pond when a stone is thrown in, but the wave lasts for no more than two vibrations. It starts where the membrane begins to be resonant, rises to a maximum and then rapidly decreases to zero. There is thus a place along the basilar membrane where the wave has maximum vibration for each frequency (Fig. 11). The length varies with frequency; not surprisingly, it is longer for low frequencies than high ones. But whatever its length, the wave running along the membrane produced by a pure tone is believed to be in the form of Fig. 12A.

As the wave travels along, any point on the membrane vibrates up and down in time in the form of the pure tone which is causing it (Fig. 12B). A frame, and its hair cell at that point, will be rocked backwards and forwards at the same rate per second as the pure tone. The hairs on the cell are thus regularly distorted by that pattern of movement. If the cell could fire a series of electrical pulses which corresponded to that pattern into its nerve fibre, that would signal the frequency of the pure tone. And that would happen with each of the hair cells wherever the membrane is resonating.

The wave (Fig. 12A) is actually a very small vibration indeed, and it was believed for many years that at the lowest sound levels which can be detected, the hair cells were signalling membrane movements of only a few millionths of a millimetre. Indeed that the ear can detect very faint sound has been described as a miracle, and when one looks at the mechanical system shown in Fig. 10B one is not even prepared to believe in miracles. What complicates things more is that the highest sound level we can tolerate is a million times bigger than that. If the delicate hairs were sensitive enough to detect the minute movements, they would certainly be irreparably damaged by the large ones. Obviously they are not.

A very recent discovery, which is still being explored, provides an answer which seems to solve this long-standing puzzle about the machinery. Attached to the other side of each triangular frame, there is a triplet of cells (Fig. 10A and

Fig. 12. (*cont.*)

given frequency resonates the same length of membrane XY, but the excursion of the wave is greater the larger the sound vibration; this is shown as increasing sizes of the envelope. Whether any hair cell fires depends on the movement it experiences at its position. If the line (T) represents a threshold, beyond which movement fires cells, more and more cells either side of the maximum fire as the envelope increases in size as indicated by the arrow (L). This signals 'loudness' as explained in Chapter 10.

C). These cells are also extremely sensitive to any movement of the membrane. They have no nerves through which they can signal anything to the brain. When these cells detect a membrane movement, they expand in length and then contract again. Let us for the moment assume they do so in the pattern of the membrane vibration. They will rock the tiny frames to which both kinds of cell are attached, but by a fixed amount, the amount by which they expand and contract. The hairs on each single hair cell would then experience a regular repeating movement at the rate of the membrane vibration at that point, but it would be independent of the *size* of the membrane's vibration. The triplets of cells can be regarded as amplifiers of very minute vibrations, and this prevents the hair cells from having to work over a huge range of sizes of vibration: they receive a 'standardised' vibration from the triplets, at membrane vibration rate (see also Appendix 11).

This remarkable mechanism was produced millions of years ago, presumably because the automatic processors need to obtain the rate of vibration of the membrane at the position of each hair cell along it, regardless of how big the vibration there is. And if the rates of vibration per second are passed on through the processors to the cortex as corresponding rates of pulses per second, then that could, for example, signal the rates of vibration of harmonics. But if this method of indicating rate eliminates the size of vibrations, and the size corresponds with loudness, the system must have a means of indicating that as well. If it has a different way of indicating the size, then we have an explanation for a valuable characteristic of musical sound, that pitch is independent of loudness. Why the processors need the ear to work this way we shall discover in Chapter 10.

9.7 Signalling loudness

Loudness is signalled as it has been from the very beginnings of hearing: by the number of hair cells signalling. The form of the resonant wave provides a way in which the loudness of a pure tone or harmonic can so be indicated (Appendix 9 and 11). Although in principle the whole length of membrane which resonates to a pure tone does vibrate as in Fig. 12A, if the sound is very small only the region of maximum vibration vibrates sufficiently to be detected by the triple row of cells, and only the small number of hair cells that they vibrate send pulses along their nerves indicating the rate of vibration (Fig. 12C). As a pure tone increases in size, more and more cells on either side of the maximum region of resonance detect the vibration; more and more hair cells signal that vibration. At any level of sound we use in music, each harmonic will vibrate several hair cells. The loudness of a harmonic may be indicated by the number of hair cells signalling its rate of vibration. The differences in the loudnesses of component harmonics, which creates the tone sensation, could be signalled by the numbers of different groups of cells vibrated at the places where the harmonics have been distributed. But the signals are assessed as a whole, for tone and loudness are each a single sensation.

Loudness is discussed again in Chapter 10. We need next to see how a pure tone's rate of vibration can be signalled as a series of nerve impulses. This will provide answers to several of the questions raised in previous chapters.

9.8 The hair cell's limitation

The possibility we have envisaged so far is of a system in which a hair cell might produce one pulse per vibration of the membrane; for example, a pure tone of 500Hz repeats its vibration every 2 milliseconds (thousandths of a second which I abbreviate to *msec*). We call the duration of one vibration the *period* of the pure tone. And if a pure tone caused 500 pulses a second, they would be spaced at intervals of 2msec, and we can use the same term, period, for the time between two pulses. A series of regularly spaced pulses with a period of 10msec would represent the repetition rate of a pure tone of 100Hz vibrating the membrane. Our models of pitch, intervals and polyphonic music have been in terms of repetition rates throughout Chapters 6 and 7. And if a stream of such regularly spaced pulses could travel through the auditory pathway, and provide the processors and cortex with pulses at exactly the rate of vibration of the pure tone, the system would be simplicity itself.

Oh! that living systems were so simple. The hair cell certainly fires because it is repeatedly distorted. The snag is produced by the biggest difference between nerves (and therefore brains) and computers. Computers can send pulses at millions per second. But when a hair cell is distorted sufficiently to fire a pulse, it has to recover before it can fire another one. It is a bit like a lavatory flush; it has to recharge before it will work again, and no amount of operating the handle will make it work until it has. All nerves are like that too. Once they have transmitted a pulse they have to recharge before they will accept and transmit another pulse; and there will certainly be a limit to the rate at which the triplet cells can expand and recover. The shortest recovery time of any nerve is just less than a thousandth of a second: 1msec. This means that the fastest pulse rate that can be sent along any nerve is just over 1000 pulses per second. It would not help if the hair cells or triplet cells could operate at a faster rate than 1000 times per second, because no nerve could transmit pulses at faster rates.

So the highest steady-sound frequency which could be signalled on a one-to-one basis of one pulse per vibration, would be about 1000Hz, the pitch-frequency of C just above treble staff. We can hear sounds at much higher frequencies. Why did our ancestors need to? There must have been real advantage in signalling higher frequencies using nerves limited to 1000 pulses a second. They found a very effective way of doing it, because our best pitch sensitivity is in the 2–4kHz range, and the cat's hearing is best an octave higher than us.

The high-frequency sound problem can be put very simply. If the membrane vibrates 1000 times a second, the period is 1msec. If a hair cell there fires a pulse, it will just have recovered in one millisecond and be able to send the next pulse, producing 1000 pulses a second, the maximum rate. But if the membrane vibrates 2000 times a second, the period is 0.5msec; two periods are 1msec. A hair cell will miss one vibration while it is recovering, and fire on the next. It also will send 1000 pulses a second. That would confuse the processors and cortex completely.

Of course it would be possible that a different method is used above 1kHz, but it is common experience that if we start with a pure tone of 16Hz and raise the frequency continuously, there is no particular change in the sensation around 1kHz. The pitch sensation rises completely smoothly up to 9kHz or however high we can hear. The only change in quality is very gradual, from being an 'ooo' sound to an 'eee' sound. (Compare the tone of vowel sounds in Chapter 5.) That would suggest that in whatever way the period of the sound is signalled, it is the same below and above 1kHz.

9.9 The discovery of the sound code

A solution to the problem is offered by some technically superb research. One group of scientists managed to record the form of the pulse streams in individual minute nerve fibres from hair cells of the cat when pure tones of different frequency were sounded in the ear; the other group carried out very similar experiments with the spider monkey. The findings were almost identical. The structure of the cochlea of both animals is very similar indeed to ours. Further details are given in Appendix 12.

Three simple tricks enable a hair cell to code the period of the sound in spaced pulses, using the same system of coding, so that they can represent only one frequency throughout the hearing range despite being limited to 1000 pulses a second. From the recordings of the pulse streams produced by pure tones we can deduce that, firstly, a hair cell never fires more than once during one vibration, however slow the vibration may be. Secondly, it fires only during one 'half' of the vibration. This creates a little semantic problem, because the membrane moves up and down, but the cochlea is a spiral of about two-and-a-half turns and your guess is as good as mine as to which is up and which down, but it won't make any difference as long as we stick to one of them, so lets call it down. Figures 13 and 14 are models which illustrate what appears to be happening (and that is why only one half of the envelope of vibration has been drawn in Fig. 12C; that represents the 'down' half of the travelling wave).

The third trick seems peculiar until we see the full implication; the hair cell does not fire on every vibration, whatever the frequency. Having fired during the down part of one vibration, it may fire during the next vibration it can, or it may miss that and fire during the next, or it may miss a few vibrations before it fires again. Which vibration causes it to fire appears to be random. I don't know, and I don't know if anyone knows, whether firing during only some of the vibrations is due to a property of the hair cells or more probably the triplet cells which transfer the vibration. And it does not matter so far as we are concerned.

Figure 15 shows what happens with a pure tone of 100Hz. The cells are

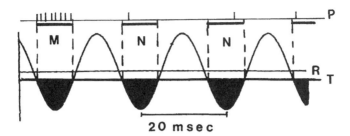

Fig. 13. Model of a hair cell detecting a 50Hz pure tone. If T is the threshold beyond which the cell may fire, it will be so distorted for several milliseconds per cycle (black area). If it were to fire as soon as it passed the threshold and again as soon as it recovered (M), it would produce several pulses (P) in each cycle. As explained in the text, a hair cell fires only one pulse per cycle (N) during the time it is distorted, and may not fire at all during some cycles.

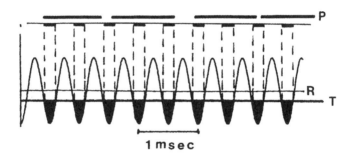

Fig. 14. Model of a hair cell detecting 2000Hz. The window during which a hair cell is distorted beyond its firing threshold is very short. Since a pulse lasts for 1msec, it extends over two cycles and the cell could, at most, produce pulses at half the rate of the vibration. Labels as in Fig. 13.

Fig. 15. Pulses from a hair cell vibrated at 100Hz. The bars (S) are spaced at 10msec intervals, the period of the sound. + = a pulse.

Fig. 16. Pulses from a hair cell vibrated at 1.25kHz. The bars are spaced at 0.8msec intervals, the period of the sound. + = a pulse. I = the number of periods between pulses.

distorted every hundredth of a second – every 10msec. The hair cell only fires on some of the distortions. The diagram shows the principle of how the pulses relate to the period of the sound.

The pulses are spaced at the period of the sound, or at two periods or at three periods. The period of 100Hz is 10msec. The pulses are spaced at 10msec, or with a gap of 20msec or a gap of 30msec. This mix of spacings cannot be produced by any other frequency. For example, a 20msec gap could be one period of 50Hz. But a 50Hz vibration cannot produce a period of 30msec. It can only produce periods of 20, 40 and 60msec.

In principle, the period and thus the sound frequency could be observed from the output of one hair cell, but one pulse train, consisting in this example of pulses spaced at one, two or three periods (as they are in the recordings from an animal ear receiving 100Hz), would have to be scanned for many pulses to be sure of what the period is. Of course, the hearing system does not 'determine the period'; that's our way of looking at the problem. Streams of pulses spaced like that cause the sensation of a pure tone of 100Hz. They signal the repetition rate of the sound causing them.

However, the real elegance of the system appears with vibrations above 1kHz (Fig. 16). Random firing also produces pulse streams which can only signal one frequency, with nerves limited to 1000 pulses per second. Between 1 and 2kHz, pulses cannot occur on every vibration, and so cannot be spaced at the period; but pulses spaced at a mixture of 2, 3, 4, 5 and so on periods of the sound provide the period on the same principle as below 1kHz. There need in fact only be some pulses spaced at 2 periods and some at 3 periods to show that the generating period must be the difference between 3 and 2 periods. That difference can only be produced by one particular sound frequency between 1 and 2kHz. There are longer gaps in these pulse trains: gaps of 4, 5 and 6 periods, but the period of the vibration must be the difference between 4 and 3 periods or 5 and 4 periods. It cannot be any other frequency. If the sound is above 2kHz, the intervals between pulses would be 3, 4, 5, 6 and so on periods, and the actual period will be 4–3, 5–4, 6–5 periods and so on. We don't imagine a pocket calculator in the cortex subtracting periods. Pulse trains consisting of multiples of the period in that fashion can convey the period of only one frequency of pure tone, and that causes the sensation of that pure tone in the cortex. Randomised firing like this is a remarkably simple way of enabling hair cells to use the same method of signalling period throughout the hearing range with no discontinuity at 1000Hz, the maximum firing rate of nerves. An authority on deciphering codes to whom I showed the system said, 'Beautiful. I wish I'd thought of it myself.'

Based on this, we can now look in principle at two important features of musical sound, our accuracy in discriminating pitches, and how rapidly we can hear that sounds have pitch.

9.10 Pitch discrimination

In order to explain the system, Figs 15 and 16 in the previous section have shown the pulses spaced in exact multiples of the period. Biological systems don't have that kind of precision. The recordings of the pulses in the nerve fibres from the hair cells which indicate that a hair cell is fired only once during the 'down' part of one vibration, also show that it may fire at any time within the 'down' part of the vibration. Figure 13 shows that at 50Hz, there is a window of several milliseconds of the 'down' part of a vibration during which a hair cell can be fired if it is vibrated by the triplet of cells. The recordings of streams of pulses from lower-frequency pure tones show that if firing happens, it is only somewhere within that 'down' window; only on average it is in the middle of the window. The spacing of the pulses indicates the period, but it is not signalled precisely. If we listen in succession to two pure tones which are actually different in frequency, and say when we can hear that the pitches are just different, we can only distinguish with certainty about twenty different pitches between 31 and 62Hz, the pitch-frequencies of the bottom octave of the double bass. Our pitch accuracy with pure tones is worse than half a semitone down there; we have to rely on the other harmonics contributing to the pitch-frequency of real instrument sound, to obtain it any more accurately.

With a pure tone of 500Hz, the 'down' half of one vibration lasts for only one millisecond and the time during which a hair cell can be fired is less than that. The spacing of the pulses in the recordings represent periods more precisely. Correspondingly, we have much better pitch discrimination; in the octave between 500 and 1000Hz (C in treble staff to an octave above) we can distinguish 180 different pure tone pitches or roughly 15 pitches in a semitone. Above 1000Hz, the part of each vibration when a hair cell can be distorted is very short indeed. At 2kHz it is less than 0.25msec (Fig. 14). The spacing of the pulses indicates the periods very accurately. Around 2kHz, we can distinguish about 240 separate pure tone pitches in an octave: about one twentieth of a semitone. It's the best pitch discrimination we have and a remarkable achievement for a biological system, but it is only accurate to five cents in a constant-frequency pure tone.

Our ability to sense pitch can never be better than the precision with which the hair cells can supply the spacing of pulses. That applies to pure tones, harmonics or any other sounds. If the vibration rates of harmonics are not quite constant, if there is timbre, that variation will be included in the spacing of the pulses too, as a variation in the periods. Our pitch discrimination will not then be as precise. Since the accuracy with which the system signals 31–62Hz is poor, any timbre variation in the sound will be difficult to observe in the fundamental of low double bass, and will only be created by higher harmonics. The higher the pitch, the more accurately the pulses correspond with the periodicity, so the greater will be the contribution of timbre to the sensation. Likewise, pitch vibrato, which is a purposely introduced frequency variation, is relatively ineffective with low bass notes, but on high notes vibrato can be very obtrusive, because of the precision with which the sound is signalled.

9.11 Minimum duration for pitch

Although theoretically we could obtain the pitch of a pure tone from the pulses from a single hair cell, there is ample evidence to show that we do not. For how long must we listen to a sound which is a pure tone, in order to get its pitch sensation? The tests have to be carried out very carefully; I've done a lot of them. If we hear a sequence of very short sounds, each actually consisting of a small number of vibrations of a pure tone, and each slightly longer than the previous one, we hear a series of very short noises, and then suddenly when the sound is long enough, we hear pitch, the pitch of the pure tone. We can detect that there is a noise, but we need a minimum number of sound vibrations to receive pitch.

With a pure tone of 100Hz, we have to hear about three vibrations (about 30msec). We obviously can't get that from one pulse train. The period is 10msec, and the first two pulses might be spaced at three periods: 30msec. With a pure tone of 2kHz, period 0.5msec, we have to hear getting on for 20 vibrations (about 10msec) before we obtain pitch. The longest gap between pulses at 2kHz may be as much as 16 periods: 8msec (Rose *et al.*, 1967). That would not give us the period either (Appendix 13). But when the sound is too short to provide a pitch sensation we do get a noise. The most primitive property of a hearing system is to distinguish a real event from the accidental firing of sensors, and unless a few hair cells fired as soon as the pure tone vibrated the basilar membrane we would not get any sensation.

The minimum duration to hear pitch does get very slightly shorter with louder sound and part of the explanation is that the membrane resonates and when a resonance receives an appropriate vibration, it does not vibrate to its full extent instantaneously. It may require three or four vibrations to build up and come into equilibrium with the sound, and may do so faster with louder sound. One cannot experiment with very loud sounds. With a sudden very loud sound, the equipment resonates, the outer-ear tube resonates, and one does not know how long the sound actually lasts. The build-up of the resonance is important with transients and is considered briefly in Chapter 10.

9.12 Producing a pure-tone sensation

A pure-tone sensation occurs when enough nerves provide pulses with the periodicity of the tone. Figures 17 and 18 show examples of the principle for a 250Hz and a 2.1kHz pure tone. There are caveats about the diagrams. Figure 17 does not indicate the variation in firing during the 'down' part of such a slow vibration. In order to demonstrate the periodicity, both diagrams show the pulses in the different nerves to be synchronised. They are not synchronised when the pulse trains go into the auditory pathway. As the wave runs along the membrane (Fig. 12A) the 'down' part causing the firing, runs in sequence past

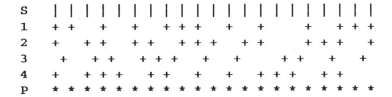

Fig. 17. Multiple pulse trains. Four pulse trains from four hair cells, 1–4, signalling a 250Hz pure tone, between them provide at least one pulse (*) at the periodicity of the sound.

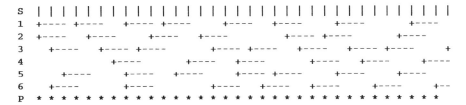

Fig. 18. Multiple pulse trains. Pulse trains from six hair cells vibrated at 2.1kHz. + - - - - indicates the duration of a pulse, 1msec. The gaps shown between pulses are up to eight periods; in Rose's measurements a few may be up to sixteen periods. Other details as for Fig. 17.

the hair cells of the region which is vibrated. And Fig. 12A suggests that if the pure tone is large, there could even be two 'down' parts of the wave traversing the hair cells at any one time. There could then be hair cells in two parts of the membrane sending pulses signalling a pure tone. The continuous sensation of a pure tone requires the continuous input of pulses signalling its period, in several nerves (see also Appendix 14). Since those pulses can be generated anywhere along the resonating length, and depend on which cells actually fire, the system does not appear able to indicate when the sound wave reaching the eardrum is going 'up' and when it is going 'down' (in physical terms, it appears unable to signal the phase of the sound). Of course pure tones and steady-pitched sounds didn't exist when this system was evolved, so there is no reason why it should.

What form are the pulse trains in when they emerge from the processors and enter the cortex to create sensations? We don't know, and looking for a needle in a haystack is much easier than putting one in a cortex and finding useful evidence. It is tempting to believe that the pulse trains entering the cortex are similar to the pulse trains generated by the ear. Since the nerves in the cortex are also limited to 1000 pulses a second, one cannot imagine in what other way they can represent pitches as accurately as they do, except by using similar pulse codes in several nerves. And since the accumulated evidence suggests that the hearing system is automatic and delivers similar sensations to everyone, the suggestion is not unreasonable.

9.13 Signalling sound with harmonics

We can now look in principle at various important musical phenomena in the light of the way in which the ears code sounds as pulses providing the period, the repetition rate of the sound. A different length of the basilar membrane resonates at different pure-tone frequencies. Such evidence as we have suggests that over the range from about 200Hz to about 4kHz, pitch-frequencies from G in the bass staff to two-and-a-half octaves above treble staff, the positions of maximum vibration of the membrane to pure tones are roughly spaced in octaves: pure tone frequencies an octave apart are equally spaced 'like a piano keyboard' though it has nothing to do with our perceiving octaves, or equal-temperament or anything of that kind (Fig. 11). Let us suppose to start with that each harmonic did vibrate a small independent piece of the basilar membrane. The hair cells under each little length would produce pulse codes each providing the periodicity of its harmonic. And if the frequencies are in simple multiples of the fundamental: $\times 2$, $\times 3$, $\times 4$, $\times 5$ and so on, the periods will be similarly simply related. They will be spaced at a half, a third, quarter, fifth and so on that of the fundamental (Fig. 19). Those are nice simple pattern matches in the pulse codes, and all those streams going through the processors into the cortex could produce a sensation which has, by our judgement of tone, good harmonicity. And if the sound has poor harmonicity, and the harmonics are not exact multiples of the fundamental, the pulse codes will not match well, and that could be why we get a sensation of different, less agreeable tone.

Fig. 19. Periodicity and harmonics. The relationship in time of the periodicity of the fundamental (F) with the next three harmonics (2–4) of a sound with ideal harmonicity.

There is a large snag in such a simple idea. If the ear were generating, for example, eight separate sets of pulse trains representing eight harmonics, why do we not get a sensation of eight harmonics? We don't. We get one pitch, and it has a single tone which, from all the evidence in Chapter 4 and elsewhere, is an amalgamation of all the harmonics. Music depends absolutely on our getting one pitch. We all experience edgy sharp major thirds; we all get the sensations of hiss and buzz from the higher harmonics of any instrumental sound. If two people discuss anything about music, it is assumed that they start with having the same sensations, whatever their opinions of the sensations are. If hearers get the same sensations from sounds with similar harmonics, then the pulse codes must be in a form which creates sensations in that form.

The answer is two-fold. The resonant lengths of membrane vibrated by pure

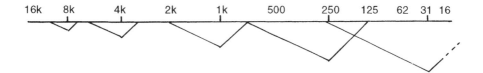

| 16k | 8k | 4k | 2k | 1k | 500 | 250 | 125 | 62 | 31 | 16 |

Fig. 20. Basilar membrane resonant lengths. Compare Figs 11 and 12. By correlating the few direct measurements made by Bekesy (1960), the frequencies which produce hiss and buzz when their lengths overlap (Beament 1997), and the range of frequencies over which individual hair cells signal (Palmer and Evans, 1995), one can obtain an estimate of the length of the basilar membrane resonant at different frequencies (see Appendix 17). The distance between the maxima of, for example, 1 and 2kHz is roughly 3.5–4.0mm, covering about 700 signalling hair cells. These estimates are used to construct the models of harmonics in Figs 21 and 23–5.

tones over the range of 250Hz to 1kHz or more, are so long that they probably have to be two octaves apart before they do not overlap at all. The fundamental and fourth harmonic are two octaves apart. And as one can see in Fig. 4, the intervals between successive harmonics get smaller as one ascends the series, so that the vibrating lengths will likewise be closer together and overlap more and more along the harmonic series. To a certain extent this is offset by the smaller resonant length as frequency gets higher (Fig. 20; see Appendix 14).

We can now examine in principle what happens on the basilar membrane when we hear a sound with a series of harmonics. As in Fig. 12, the region vibrated by a pure tone is shown by a triangle, which represents how far the 'down' part of the vibration extends as it runs past the hair cells and distorts them.

Figure 21A represents two pure tones an octave apart, as are the fundamental and its second harmonic; these are always the most widely spaced of any adjacent pair of harmonics in a sound with a complete series of harmonics. There is a length of membrane producing the periodicity of the fundamental, say 250Hz (roughly middle C); another length is producing the periodicity of the second harmonic, 500Hz. But there is an overlapping region being vibrated at both rates. If the two frequencies are in the ratio of 1:2, the membrane may accommodate this easily, because every other vibration at 500Hz will coincide with and reinforce one of the 250Hz vibrations. Rose *et al.* (1967) did a few experiments of this kind, with two pure tones an octave apart. The pulses in the nerve from a hair cell so vibrated showed the periodicity of the higher frequency with the periodicity of the lower frequency superimposed on it; it signals a strong pulse rate period of the fundamental combined with a pulse rate period of the second harmonic (Fig. 22).

But the second harmonic of a typical musical note has the third harmonic on the other side of it, and overlaps it too (Fig. 21B). The overlap region is vibrated at 500 and 750Hz. If the harmonicity is good they reinforce each other 250 times

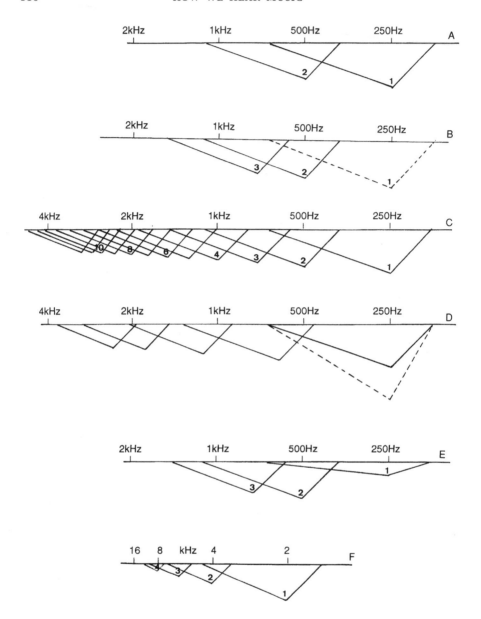

Fig. 21. Vibration of the basilar membrane to pure tone harmonics. A: A long length of membrane is vibrated by the fundamental (1) of 250Hz (middle C). A significant length is simultaneously vibrated by it and its second harmonic (2) of 500Hz. B: second and third harmonics. Unless the fundamental (dotted lines) is missing, no part of the membrane is vibrated by the second harmonic alone. C: The first twelve harmonics of a 250Hz pitch-frequency note assuming a uniform decrease in their sizes. By the seventh harmonic, the overlap produces hiss. From harmonic 11 upwards, only hiss is produced. The only harmonic signalled on its own is the pitch-frequency. The rest produced the

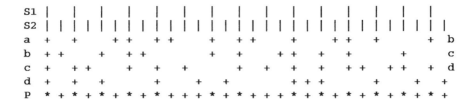

Fig. 22. Model of pulses from two pure tones. One hair cell is vibrated simultaneously by 100Hz (S1) and 200Hz (S2). Other labels as in Figs 15–19. Line (b) is a continuation of line (a) and so on. If this pulse stream is analysed as shown in Rose's recordings, it has a majority of pulses with a period of 100 per second (*) interspersed with pulses (-) at 200 per second (T).

a second too, at the period of the fundamental. The point of maximum vibration of the third harmonic is closer to that of the second harmonic (see Fig. 4), so that the third harmonic overlaps the second more. We may question whether any part of the second harmonic is vibrating free of the first or third harmonic. Similar conditions prevail as we ascend the harmonic series. The third harmonic, 750Hz is overlapped by lengths of the second and fourth. The overlap of 750 with 1000Hz may also have hair cells producing a periodicity of 250 pulses a second where the vibrations coincide, as well as of 750 and 1000Hz. Of course this does not go on for ever; as the harmonics get closer, the membrane is trying to vibrate at three, four and perhaps more harmonic rates at the same time (see Fig. 21C and the following section).

9.14 The generation of hiss and buzz

Figure 21C shows in principle how the higher harmonics of a note with 250Hz fundamental become more and more crowded and overlapping. At some point, the basilar membrane cannot vibrate accurately simultaneously at several frequencies and its movement becomes irregular; irregular vibrations produce irregularly spaced pulses from hair cells, and that creates a noise sensation. That can be demonstrated more simply. Two simultaneous pure tones with frequencies in the ratio of 7:8, for example, 1750 and 2000Hz produce some hiss; that is the ratio of the seventh and eighth harmonics of 250Hz. The hiss increases as the tones get closer, and less of the tones can be heard; that means more random pulses and less pulses signalling the frequencies are produced,

Fig. 21. (*cont.*)
amalgamated tone. D: The first five harmonics of a xylophone note. The pitch-frequency harmonic would probably be comparatively large (dotted). The overlap is small and part of the second and third harmonics are signalled on their own; they can sometimes be heard. E: The first three harmonics of a note with a very weak fundamental. F: The harmonics of a musically high note (C two octaves above treble staff). Only the first two or three harmonics are significant.

until when the two pure tones are in the ratio 11 : 12, equivalent to the eleventh and twelfth harmonics of the note, they overlap so much that they obliterate each other and all we hear is hiss or roughness, the sensation described in Section 4.16. If a pair of pure tones produce hiss like that on their own, they certainly will in instrument sounds; it may happen lower down the series as the harmonics become crowded. From using pairs of pure tones as an indicator, we can say with certainty that notes from about violin open G in the bass staff, to the top of the treble staff, will produce hiss and noise from at least the seventh harmonic upwards, and from the eleventh harmonic upwards only noise is produced (Beament, 1997). We saw in earlier chapters that harmonics played no direct part in the original selection of the pitches of scales; if the harmonics are reasonably similar in size, the higher ones cannot contribute anything useful about their vibrations at all. One young composer who questioned in a national newspaper what a professor of biology (myself) knew about music, wrote that modern music was 'an infinity of harmonics'. They would make meaningless chaotic noise; further comment would be superfluous.

9.15 Low and high frequencies

The problem the ear has with very low frequencies is that the resonant lengths get longer and longer, but the basilar membrane does not. One could draw the conclusion that if hearing low frequencies with the discrimination we have at higher ones, had been valuable for survival, natural selection could have produced a basilar membrane twice as long. The overlapping of the resonances of low harmonics is demonstrated by pairs of pure tones. Some buzz is produced even by a pair of tones corresponding to the second and third harmonics of cello open C, and by the overlapping of all its harmonics up to the tenth harmonic, beyond which only buzz is produced. Buzz is the sensation of low random noise, just as hiss is produced by high-frequency random noise. Over the bottom octave of the double bass, even the fundamental and second harmonic produce some buzz, but we can hear something of the periodicity of the harmonics up to the sixteenth harmonic.

Hiss and buzz are generated by our ears. There is no sign of noise in the sound in the air from bass instruments. And if one of a pair of pure tones which produce hiss or buzz when heard together is fed into either ear, there is no roughness sensation; the two tones have to interfere on the same basilar membrane.

Buzz is a characteristic of any low note with a complete series of harmonics. We hear it from the contra-bassoon, the bass saxophone and the reed pipes of organs. We do not hear it from a bass clarinet, because the sound consists mainly of only harmonics 1, 3, 5, 7, 9 and so on; they are twice as widely spaced on the basilar membrane and do not overlap sufficiently to produce random noise until the twelfth harmonic or higher. Clarinet sound is very 'clean' because of this. There is very little hiss from higher sounds with few harmonics, such as from muted strings. Their harmonics do not extend far enough to generate it. But there is still some buzz from the bottom octave of muted double bass. Buzz is usually attributed to the instrument. People have said that it must be due to the way the reeds of the bass wind instruments vibrate. I've even been asked how one could make a double bass that doesn't buzz. Of course, some of them have two buzzes, one produced by the harmonics and one produced by a glued joint that has opened. That can make life very difficult for a repairer when the customer says 'I can still hear a buzz.'

There are problems about how we hear very high frequencies (Section 9.10 above and Appendix 15) but they have very little musical value. There are physical difficulties in producing any pitch-frequencies above about 2kHz (three octaves above middle C) on any orthodox musical instruments. Put the other way round, our hearing and physics determined what we use in music. That has been the yardstick ever since the first page.

There are a number of things which we can discuss further, arising from the way in which the frequencies of harmonics are distributed, but Fig. 21C appears to provide an answer to the question of paramount importance.

9.16 The creation of the pitch sensation

In Fig. 21C the only harmonic which has a large length of its vibration, including its region of maximum vibration, free of any overlap with other harmonics, is the fundamental and that will be signalled on its own. That is the pitch-frequency. Because of the way frequencies are distributed, the lowest frequency in any sound made up of several frequencies, is always able to signal its periodicity by a part of its vibration, free of other frequencies. Musically speaking that applies just as much to the received pitch of percussion instruments such as the xylophone (Fig. 21D) and steel drum; the pitch is that of the lowest harmonic. Of course with pitched percussion sound, the lowest harmonic is almost always by far the biggest too. The situation also appears applicable to any noise, whether it is a word or other noise, provided it does not consist of a very few widely different sound frequencies; all noises have a generalised non-musical pitch characteristic of being 'high' or 'low'. This is discussed again in the final chapter in relation to the human voice.

That is the best explanation I can offer for the all-important sensation of pitch upon which western music absolutely depends. There are other theories which suggest that there is some sort of automatic filter or 'comb', which selects or recognises the vibration rate common to all the harmonics, that is to say, the fundamental, but they appear to fail on two grounds. Firstly, they would only work with sounds which when analysed do have reasonable harmonicity, and not those which have no harmonicity, a random collection of frequencies such as characterises natural noises. Secondly, there is no biological reason why noise of any kind such as musical notes should be presented to a cortex with a pitch characteristic. Pitch appears to arise automatically from the way sound frequencies are distributed and coded by the mammalian ear and then processed. That will happen whether the animal has a cortex which can or cannot recognise pitch!

The production of the pitch period by, for example, overlapping lengths of the second and third harmonics, explained in Section 9.15 above, provides a basis by which the harmonics can augment the pitch sensation; that is very important where instruments have weak resonance in the low part of their range, such as the bottom notes of the violin family (Fig. 21E). It is very important in the establishment of low pitch itself. The experiments with a synthesiser demonstrated that one could obtain a pitch sensation even when the fundamental's vibration was removed from a set of pure tones forming an ideal harmonic sequence (Fig. 21B). But one must interpret such experiments carefully. By removing the fundamental, a larger length of membrane is left vibrating simultaneously at the second and third harmonics' frequencies. And with a real instrument's sound, the pitch sensation is only augmented because there are large harmonics in the series next to the weak fundamental (Fig. 21E).

9.17 The creation of tone

In Fig. 21C, apart from the fundamental, all hair cells throughout the vibrating region which signal, are signalling two or more periodicities simultaneously, a mixture of the pulse rates of the harmonics. Thus the signals representing the rest of the harmonics are all mixed together, amalgamated, as it is our experience, into the single sensation of tone. Since that appears to happen (as we also saw with synthesiser experiments in Chapter 4), tone cannot be a very well defined and uniquely memorable feature, and that is also why it is so difficult to describe. The only variation in tone from sounds with good harmonicity, is produced by the relative sizes in the mix of harmonics. If a note in the treble staff has strong harmonics in the 2–4kHz region where hearing is most sensitive (amongst other things because that range is enhanced by the resonance of the outer-ear tube), the tone sensation will be strong, and the sound of the instrument will 'project', which simply means that it will be well heard in a large auditorium. If there are few harmonics, the tone will have a weak characteristic. That is typical of muted strings, and of the low notes of flutes and of the clarinet and bass clarinet, which are very easily masked by other instruments and require careful orchestral accompaniment.

If the length of the membrane vibrated by the harmonics of notes above treble staff is small, the general form of the distribution of the harmonics can be illustrated as in Fig. 21F, a compressed form of Fig. 21C. But a different factor enters into the situation. The C an octave above treble staff has a fundamental of about 2000Hz. Its fourth harmonic is 8kHz. The first two harmonics are strongly signalled. But the third and fourth harmonics are in the region where our sensitivity drops dramatically, for we hear very little of harmonics above about 6kHz at all. As a result, very high notes on instruments are clear and relatively free of noise; their pitch frequency occurs where our hearing is most sensitive to both pitch and loudness. But they do not have any very marked tonal character.

We now see what happens if the high frequencies are boosted on radio and in CD recordings. Any change of balance of the sound between high and low frequencies changes the tone, and any increase in the size of high frequencies simply makes our ears generate more noise.

9.18 Simple simultaneous intervals

The simplest interval is a precise octave. Provided the two notes have good harmonicity, each of the upper note's harmonics produces a vibrating length coinciding with every other vibrating length of the lower note (Fig. 23A; see Fig. 4). This does not alter the overall form of vibration of the lower note. All it will do is increase the size of the matching vibrations and therefore the number of pulse streams signalling their vibration rates, in the amalgamated tone envelope. The tone will be richer, the projection of the sound greater, because it will inevitably increase the strength of higher harmonics. It does not affect the signalling of the pitch from the fundamental of the lower note. It is a standard rule in orchestration that one can double at the octave above without changing the received pitch, but one does not double at the octave below, for that drops the pitch an octave.

This reveals yet another bit of confused thinking by some musicians. A simultaneous octave is a special condition. It is, as the following paragraphs show, unlike any other interval. But a consecutive octave is no different to any other pair of consecutive pitches. They are just related pitches. It is therefore illogical to consider notes which are octaves of each other as the 'same note' in such completely artificial devices as a tone row.

Any other simultaneous interval produces a situation which differs from an octave. The harmonics of a fifth [CG] are illustrated in Fig. 23B. The vibrating length of the fundamental of G occurs with a maximum vibration between the first and second harmonics of C. (One can see all the coincidences and lack of them by looking at the pitches of harmonics in notation form in the manner of Fig. 4.) The fundamental of G reduces the length of membrane signalling the pitch-frequency of the C alone. And as the diagram shows, the two sets of harmonics which coincide if the interval is reasonably accurate with good harmonicity sounds, are octaves of the upper note. The rest of the two sets of harmonics are overlapping as closely as are pairs of pure tones in the ratio of seconds (7:8) or semitones (15:16), noise or complete roughness. The reinforced vibrations are those which identify the upper note. A simultaneous interval is a unified sensation, with a unified tone, which we usually tend to identify by the pitch of the upper note, not the lower one as we do with the octave.

We can apply similar argument to the interval of the fourth and major third. Because there is less distance between the fundamentals of a fourth or major third, the length of membrane vibrating at the fundamental of the lower of the

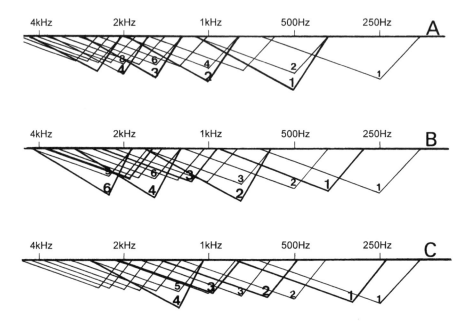

Fig. 23. Vibration of the basilar membrane to intervals. Simultaneous intervals with the lower note pitch-frequency 250Hz. A: An octave. The harmonics of the higher note (heavier lines and figures) reinforce every other harmonic of the lower note. The pitch-frequency of 250Hz is still clearly signalled. The tone is changed. Harmonics beyond the fifth one of the upper note would be in the overlapping region producing hiss. B: A fifth. The length of membrane signalling the 250Hz pitch-frequency is smaller. The reinforced harmonics are octaves of the *higher* note and may obliterate higher harmonics of the lower note. Overlap in the 2–4kHz region is extensive. C: A major third. The region signalling the pitch-frequency of the lower note alone is further reduced. Harmonics 4 and 8 of the upper note are reinforced and will tend to obliterate smaller harmonics. The upperer note's harmonics between 4 and 8 are not shown but their overlap with the lower note's harmonics will obliterate most of the information about the harmonics of both in this region.

two notes is even smaller (Fig. 23 C). In the fourth [CF] there is reinforcement of the fourth harmonic of C with the third of F. In the major third [CE], the fifth of C reinforces the fourth of E. Otherwise, as a comparison of the two sets of harmonics of these intervals in notation form shows, the rest of the harmonics are as close as seconds or semitones.

9.19 Other simultaneous intervals

If we consider other intervals in a similar way, which can be done simply by comparing the harmonics of the two notes in notation form, we can see that any possibility of a match between harmonics becomes of less and less value. For example the minor third [CE♭] provides the fundamental and second harmonic of both notes. Above that with the exception of the fifth/sixth harmonic they have in common (G), all the harmonics considered as pairs have either a ratio of roughly 7:8 or 15:16: noisy, or rough and self-eliminating. To that we have to add that from the seventh harmonic upwards of each of the component notes, noise appears, and combinations of the two sets of harmonics will effectively eliminate any contribution to the interval sensations. But there is no problem in producing the sensation of a minor third on orthodox instruments or recognising it. A similar argument can be applied to the major sixth [CA]. And the simultaneous augmented fourth has all the harmonics in completely rough appositions; when combined, they will make little contribution to the determination of the interval at all. It always has been a troublesome interval to the inexperienced.

With a simultaneous second [CD] or minor seventh [CB♭] every harmonic of one will be in the ratio of 7:8 with the corresponding harmonic of the other, and produce indistinct signals, while every harmonic of simultaneous semitones [CC♯] and major sevenths [CB] produce roughness. In Chapters 3 and 5 I said that the simultaneous second was not accepted for a long time, and that generally speaking, simultaneous semitones [CC♯] still are not. Even the fundamentals of a simultaneous semitone overlap almost completely. We can learn to recognise one as a noise made by an instrument like a piano; we don't recognise it as an interval in the way that we recognise a fifth. So-called clusters: groups of simultaneous semitones such as are used, for example, by Ives, are noises; they can be effective but may be regarded as roughly pitched percussion sounds.

The previous paragraph refers to pairs of pitches from the middle of the bass staff upwards. But even the second and third harmonics of notes below bass staff overlap and produce buzz. Those two harmonics are in the ratio of a fifth. Fundamentals more closely spaced than that, and their harmonics, produce a large amount of buzz. Hence closely spaced intervals of bass notes sound very muddy. The standard text-book advice on scoring polyphonic music, to use large intervals in the bass, has a sensible basis in the mechanism of the ear. There is a further discussion of very low frequency in Appendix 17.

9.20 Chords

If we extend these conclusions about simultaneous intervals to chords, the more notes which are heard together, the more the higher harmonics interfere with one another, and recognising a chord depends primarily upon the repetition rates of the pitch-frequencies, in other words the pitch patterns as described in Chapter 6. The sensation which is obtained from all the overlapping harmonics is a single amalgamated tone envelope. A chord has a single tone. We may be able to learn to recognise it as a bowed string chord or a brass chord or a wind chord, but the major part of the recognition of the instruments is done by using the starting transients.

9.21 Harmonicity

Harmonicity plays two roles. Since the lower harmonics contribute to the pitch sensation, if the harmonics of one note are not in reasonably exact ratios, the pitch frequency they generate by their interaction where they overlap on the basilar membrane will not match the fundamental pitch rate, and therefore the accuracy of determining pitch will be blurred; as we said earlier, notes with poor harmonicity have a wide pitch-band. Notes with timbre have a slightly wider pitch-band too. It is fascinating that musicians like one and not the other. In all probability, timbre helps to disguise slight lack of harmonicity, apart from its other valuable contributions.

Wind and brass instruments were improved over a couple of centuries by instrument makers by trial and error, using their hearing to judge the tone of individual notes. If the amalgamated pulse rates generated by the harmonics do not form a good simple-ratio series, musicians often do not like the tone. We cannot say the tone is poor, because the judgement of tone is a matter of taste. We can say that neither the tone nor the pitch will be clear, and that musicians like tone with good harmonicity because it enables them to hear themselves and others more accurately and to make more accurate intervals. And it helps to make woodwind and especially brass instruments more user-friendly, because the 'harmonic series' of the brass instruments will correspond more closely to the required intervals.

The difference between tone with good and poor harmonicity can readily be experienced today because bowed gut and the best polymer-cored strings have almost perfect harmonicity whereas steel cored strings do not. Musicians can easily distinguish the two in blind tests. Steel strings make a louder noise and they make playing easier: two attitudes to music characteristic of the end of the twentieth century. However, to many people they do not make as attractive a sound.

But if the harmonicity of two instruments' notes is really poor, as it was with all the early woodwind and brass instruments, there will be a poor match between those harmonics which should coincide and therefore one will get unattractive interval sounds; it will be difficult to play intervals with any degree of accuracy. This provides strong support for the argument at the beginning of Chapter 6 that polyphonic music did not develop until sounds with reasonably good harmonicity could be used.

9.22 Conclusions about intervals

The general conclusion is that since harmonics are amalgamated in tone, we can make very little direct use of them in determining intervals, other than the simultaneous octave, and that the more notes which are heard simultaneously, the less direct help the harmonics provide. The intervals and chords are very largely determined by the pitch patterns, as we saw in Chapter 6. The patterns are provided by the way the sound is coded. We can obtain the pattern of several intervals when the match is not particularly good. A major third has a major third sensation when the ratio of the pitch-frequencies departs very considerably from an exact 4 : 5 ratio. The only thing which shows the departure is the tone, and that only works adequately if the harmonicity is good. That is apparently the main role the harmonics play.

9.23 Consecutive pitches

The weight of the argument about simultaneous intervals and chords: that using them is very largely dependent on the pitch patterns, serves to support the original proposition in Chapter 3. The inventors of pitched music remembered one pitch and compared that memory with a second pitch. It didn't matter what the tone, the harmonicity was. They did it with xylophones, and with panpipes which generate a poorly matching set of only harmonics 1, 3, 5, 7, and so on. They remembered the one periodicity clearly signalled, and they compared it with a received periodicity of a consecutive note's pitch. I do not think there is any alternative to that proposal; it is comparing pitch patterns in sequence. And pitch patterns are provided by the periodicity of the pulse streams generated. Furthermore, I strongly suspect that that is what we still do. In general, most people remember melodic lines as pitches, and pitches are single repetition rates (and see Chapter 12). I make a very tentative suggestion in Appendix 19 that we may actually be able to store melodic material in the form of single pulse streams, and recover them as pitches. And in respect of harmonics, many experiments have shown that we cannot remember tone as we have defined it, the amalgamated harmonics, in any detail. We cannot even always reliably identify what instrument is being played by the tone of the steady-pitched part of their notes, let alone the tone of individual violins.

9.24 Polyphonic music

In the context of the present enquiry, no useful purpose would be served by looking at the interaction of harmonics in more complex musical situations. When several instruments are playing, two closely adjacent similar sized small harmonics from two instruments obliterate each other in noise. Large fundamentals will be signalled wherever they are, and large harmonics from powerful instruments will swamp or obliterate small harmonics of the same (or nearly the same!) frequency. In an orchestral tutti, the fundamentals and some of the other harmonics of the brass instruments are big enough to swamp most of the harmonics of the rest of the orchestra. A lot of buzz and hiss is generated by our ears. It is part of the normal sensation. If you listen carefully when the whole orchestra is playing, you are very hard put to say how much of the contribution of, for example, the strings, is a complex mixture of pitched sounds and how much is noise. We will hear violins and flutes if they are scored above the rest of the orchestra, in the frequency range where our hearing is more sensitive. But in good orchestration the more sets of instruments that play in unison or double at the octave, the less able we are to distinguish tone; the more the sound conveys the pitches of the parts: the pitch-frequencies. There will be a lot of noise; that is all part of the sensation. We normally don't listen for the noise, and therefore we don't observe the noise, any more than we listen for the buzz of low cello when hearing a string quartet. It isn't a very loud sensation, particularly at a sensible distance from the instrument.

9.25 The general musical pitch phenomenon

Our aim has been to discover how the notes of music are presented to the cortex in such a form that they provide the sensation of pitch and convey the relationship of pitches, and how notes convey tone. It may be worth looking back at Fig. 5 and seeing that such different forms of vibrating air all have the same received pitch. A sound wave, vibrating in a particular way in the air, does not have pitch. It does not consist of a series of harmonics. It is air vibrating in a complicated repeating way. It is only because of how our basilar membrane vibrates, partially separating the sound waves into what we call harmonics, and putting the slowest of the vibrations on its own at the low-frequency end of the system, enabling the hair cells to signal it, that such a thing as pitch exists and it only exists as a sensation while we are hearing it. That is why I object to the use of the word pitch as a physical term. Physically, sounds have frequencies. Pitch would not exist unless our hearing system did. In an analogous way, light does not have 'colour'. A very small part of the huge range of electromagnetic frequencies from X-rays to long-wave radio, produce colour because eyes see that tiny range as colour. To discover what light frequencies there are in a colour, we have to use a prism. We see a single colour made up of those frequencies. We hear a single tone made of harmonics.

And in discussions of hearing, I have read all too often that the cortex 'looks' at patterns of vibrations, or patterns of nerve signals, and 'sees' relationships. I had to invent the word 'auralise' because we cannot visualise music. If the cortex observes any relationships between sound sensations it *hears* them. One may need to use visual analogies and diagrams to explain things, but we cannot see sound, either as dots or waves on oscilloscope screens or in our cortex.

As a basis of orthodox music, the idea of matching pulse rates does not appear to be more difficult to grasp than the Helmholtz theory of harmonics, and it has one feature which makes it superior to any other theories which I have seen. It uses a processing system which was already in place a million years before music was invented. We have yet to discuss why the hearing machinery distributed sound in its different frequencies, and coded sound frequency as pulses with sound periodicity. The day steady-pitched sound was produced accidentally, the system processed it and produced sensations with the pitch characteristic. The cortex did not have to learn anything, and it does not have to do so to obtain the sensations of music. Sounds with simply related pitches produce relatively simple sensations. It isn't just coincidence that the majority of unsophisticated listeners like simple tonal harmony and melodies based on the simple relationships of pentatonic and major heptatonic scales, without any idea of what they are. The relationships are there in the nerve pulse codes. From such a starting point, music was developed empirically into a very sophisticated activity using those relationships, and eventually into a modern art form which seems to have forgotten its origins.

9.26 Anomalies in pitch perception

The scientific study of pitch perception reveals some unexplained phenomena. Almost all the apparent anomalies arise from experiments using pure tones. It is only worth describing briefly a couple of these, both of which have been known for over fifty years. If someone hears a pure tone of 1kHz at a low level, and loudness is increased to a very high level, the pitch does not change perceptibly. With 8kHz, the apparent pitch may rise with increasing loudness by as much as two semitones, and with 150 Hz the pitch can fall by almost as much, but the extent to which individuals experience it varies a great deal (Wever, 1949). It doesn't happen with music, that is to say, sounds with many frequencies, and it would obviously be a disaster if it did, for the volume control of an amplifier would not work.

The more peculiar anomaly is called the mel scale. This assumes that a pure tone of 1kHz produces a 'pitch' of 1kHz. Subjects are asked to remember it and then select a pure tone which is 'an octave' above it, and then, remembering that pure tone, to find one 'an octave' above it and so on. Likewise, subjects select 'an octave' below, and continue in octave steps downwards. The pure tone frequencies of these 'octaves' are plotted against perceived pitch. The results might suggest that if we hear a pure tone of about 6.1kHz in isolation, it produces a pitch sensation of two octaves (=4kHz) above 1Hz, while a pure tone of 200Hz produces a pitch sensation of about 260Hz. In other words, the pitch of a pure tone is about a fifth out at very high frequencies, and a third out at very low ones. Houtsma (1995) in a paper fortunately untypical of most scientists who work in this area, describes all the anomalies with little attempt to suggest any explanations, nor even to try to reconcile the mel scale with the apparent pitch change with loudness, or how any of the standard pitch and loudness tables were obtained. Under normal circumstances of listening to music we never hear a single harmonic at anything like the loudness level at which there are significant pitch changes. But Houtsma adds, slightly sarcastically, that 'the mel scale never became quite as popular with musicians as the comparable sone scale for loudness'. I don't suppose the few musicians who have heard of either scale would be in the least interested, and as we shall see in the next chapter, the sone scale is utterly impractical and useless too.

The reaction of musicians to things like the mel scale is to regard any experiment with pure tones as unreliable, which would be sad because an enormous amount of valuable scientific work on hearing has depended entirely on using pure tones and modified pure tones, with no attempt to relate it to music. The alternative view, which I have taken, to which there are references in the papers cited by Moore (1995), is that we use the relationships between the pulse codes in the nerves, and I have adopted it for the simple reason that it appears to correspond with most musical phenomena over the range of frequency and loudness used in music. Music is concerned with the sounds of instruments and voices with several harmonics, and such things as the mel scale suggests do not appear to happen with those sounds.

There often appears to be an undercurrent of wishful thinking when prominence is given to phenomena like the mel scale and pitch change with loudness, because they might support the view that pitch is signalled by the position of the resonant vibration on the basilar membrane. If that were so, the lack of precision it would produce suggests that no one could tell the pitch of any pure tone to within several semitones (Appendix 14). The idea that position is used is also at odds with the mechanism of loudness discussed in the next chapter. Now if one believed that pitch is determined simply by where the basilar membrane vibrates (Appendix 15), and not by the coding of frequency, one could also believe that all scales including the pentatonic and heptatonic are completely arbitrary. Houtsma apparently believes it to be so, suggesting that older children have to learn solfeggio in order to acquire the pitches of the dodecaphonic scale. Well of course solfeggio doesn't mean that at all. It is voice training, learning to sight-read notation for singing and learning to control the voice to produce the pitches required. It is directly comparable with learning to play a violin from notation and to produce the right pitches. In both cases, the pitch required is determined by hearing, and people can play instruments and sing without learning notation at all. Indeed if children learning the violin do not fairly rapidly demonstrate that they can use their hearing to produce reasonable intervals, one suggests other activities; one might wish it had been suggested to some singers despite the amount of solfeggio to which they have been subjected. It is widely acknowledged that musical illiterates, the jazz and traditional fiddlers who play entirely 'by ear' have superb intonation. No one has ever questioned my double bass intonation, despite the mel scale.

9.27 A really aberrant hearing system

The following story emphasises how automatic a hearing system can be when there is a fault. In 1949 a freshman, son of a well-known musician, auditioned to join the Cambridge Footlights Club. He proposed to tell a series of jokes, ending with a well-known 1920s song, but he didn't have a copy of the piece. He was told that I would play the piano for his song, because I didn't need it. When he got to the Clubroom he gave me a scrap of paper with the last lines of his patter, at which I was to play an introduction, which I duly did, and he started to sing – a fifth out from the key I had provided. So I changed to his key, he switched another fifth, and we chased each other through most of the cycle. And the committee thought it was sensational – until they discovered that he couldn't sing any other way. There are a very few people who always pitch a fifth out when they try to sing, and a larger number who do so when whistling. Since this is obviously automatic, it supports the belief that the ears and fixed processing system determine the form in which the coding representing sound is delivered to the cortex. But it presents a real challenge to those who investigate hearing, to provide an explanation.

9.28 Tracking and the three-tones paradox

In Section 7.5 we described the way in which the combination of three pure tones of 200, 600 and 1000Hz could be heard either as a pitch with tone or as three separate pitches. It was used there as an example of how our hearing system, when it receives a new sensation, puts it into short-term memory and continues to track it, but cannot separate three harmonics if they are received together, and that we cannot unravel wind chords if they are ideally chorded when playing. The probable explanation of the paradox (Fig. 24) is that the distribution of the resonances on the basilar membrane is on the margin of amalgamation or separate vibrating lengths. The paradox works with headphones. If we hear the three sounds together from speakers in a room, they are differently reflected by the walls, their respective loudnesses change with each movement of the head, and hearing may then separate them into the components. It is unwise to demonstrate any acoustic experiment in a lecture theatre without first discovering whether it will work. Although one could never separate the component harmonics of an orthodox instrument's sound by differential reflections, one cannot assess its sound adequately in a typical domestic room because of the way different frequencies are reflected. The corollary of this experiment is that if a set of harmonics are much more closely spaced, as they are in the sound of normal wind or stringed instruments, we will not be able to hear them as separate entities.

The three-tones paradox is no longer a paradox, but tracking any sound as a distinct entity once it has been identified as such, is a significant feature of the hearing process. Much more important, however, is that we also track a noise we hear in the environment, and we can keep it distinct from other noises occurring at the same time. That is the problem discussed in the next chapter.

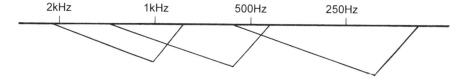

Fig. 24. The three-tones paradox. Significant lengths of membrane resonating at only one of the three frequencies, 200, 600 and 1000Hz, allow the individual components to be heard separately or amalgamated according to circumstance.

10. A sense of direction

10.1 Introduction

Why should we have a mechanism which can keep some sounds which are separate entities separate from one another? The usual explanation attributes everything to the cortex, which is a superior way of saying we don't know. The standard example always described, and said to indicate the remarkable ability of a cortex to extract one of two or more received sets of sounds, is called the cocktail-party phenomenon. It enables us to overhear a conversation far more interesting than the one we are suffering from the person who has latched onto us (and today usually with some mindless music being played at the same time). It is a bad example because it confuses three phenomena: the direction of the sound, tracking, and using our knowledge of language to interpret partially heard word noises.

If we receive noises from two different directions at once, and we could not distinguish the pulse streams which they generate at the same time in the ears, we would not know that they were separate sounds. But the sensations arrive in the cortex as separate noises, each with its direction. It may not occur to everyone that direction is just as much a sensation of a noise as is its general pitch or anything else; it is a more basic feature of the hearing sensation than any other characteristic except loudness. We can track and concentrate on either of two sets of noises, and that is a cortical activity. But no cortex can re-assemble a vast stream of pulses arising from two or more complexly changing distinct noises if it does not know how the noises are going to change instant by instant. We can learn to track the contributors to chamber music from more or less one direction, but music is a special artificial kind of sound.

Many mammals can obtain the direction of a sound source in the presence of other sounds, and they do not have advanced cortical processes to track one sound sensation from the mass of pulses coming from their ears. Neither they, nor we, have a problem with a noise that has never been heard before. Sound location is automatic, performed by processing stations before nerve pulses reach the cortex. The processors can send instructions about sound direction to eyes, neck, or rotating external ears; it is a reflex response to one noise in the presence of other noises. That is the survival purpose for which it was evolved. Most of us live in noisy environments, and receive noises from all directions. We do have awareness of their directions, though I am sure many people don't bother to observe it.

The development of direction-finding has had a profound effect on the entire nature of hearing. Its automatic processing units, which are the most recently evolved part of the auditory pathway, and, not surprisingly, the most complex part (see Fig. 25B in Appendix 8), appear to have dictated the properties of the ears, and that determined the nature of the sensations we receive. Very little of noise of survival importance does not contain a lot of sound below 1kHz. It could be signalled with a simple coding system. It would provide a satisfactory communicating system for any animal with a suitable cortex (and indeed a form of music with an upper limit of hearing of C above treble staff). So why did we and other higher mammals evolve a special coding system which signals frequencies up to 9kHz and beyond, and which provides us with our best sensitivity between about 1.5Hz and 4kHz? Why do we have the peculiar relationship between loudness and sound frequency which was described in Chapter 8? Take any characteristic of hearing discovered in black-box experiments, or critically observed in musical sensations, and ask why should it be like that; the only convincing reason we can advance is, because that is what direction-finding required. The processing units have not been elucidated in detail, but the difficulties in investigating them are immense. There is a remarkable example of what the processors can do that a cortex cannot in Section 10.6.

Direction-finding is based on two features of sound: the size of the vibrations which we characterise as loudness, and the time of receipt. We will take loudness first, then the timing problem, and finally the way they have been exploited by such artificial tricks as 'stereo' and 'surround sound'.

10.2 Signalling loudness

In almost every animal sense which has been investigated, an increase in the intensity of a stimulus is signalled by an increase in the number of pulses per second sent to the brain. The commonest mechanism is to have sensing cells which each produce more pulses per second the more they are distorted. That cannot be used if the pulse rate has to signal frequency. A second way is to have cells of different sensitivity so that as the stimulus increases, more and more cells fire. The system we have described appears to need all the cells detecting vibration to have similar sensitivity, for a resonant vibration can occur anywhere along the membrane. And since each cell only detects the vibration at its site, which is vibrated by a relatively limited range of frequencies, but a natural noise can have any number of frequencies, typically changing rapidly, the overall size of a noise, its loudness, could only be obtained by summating all the pulses, regardless of where they originate. That, it is suggested, is what the dorsal cochlear nucleus did (Appendix 9); perhaps it still does with random noises.

Once the system signalling vibration rate, described in the previous chapter, is established, the system is ideal for signalling increased vibration of music sound. As the size of a pure tone or harmonic increases, there is an increase in the number of hair cells on either side of the region of maximum vibration that experience sufficient vibration to fire (Fig. 12C). But the streams of pulses from

Hz	31	62	125	250	500	1k	2k	4k	8k	(16k)
fff	1.2	1.5	2	4	10	18	29	26	23	10
mf	1	1	2	4	9	17	27	26	23	10
mp	0.5	1	2	4	9	13	25	24	21	10
ppp	0.3	0.6	1.5	2	6	13	22	21	17	10
D.	1.9	3.7	3.7	3.7	3.5	3.4	4.1	9.2	20	96

Table 3. The number of pure tone pitches we can discriminate in a pure tone semitone at different frequencies. 250Hz = middle C. 0.5 means that we can only just discriminate pitches which are two semitones different. D is the same data for *mf* but expressed as the difference in the rates of vibration which we can detect, which remains remarkably similar up to 2kHz. Data for 16kHz for young persons who can hear that high.

one pure tone or harmonic, as illustrated in Figs 17 and 18, are statistically identical, that is to say, they all consist of random numbers of periods of the pure tone. One can add any number of further pulse trains of the same kind. They increase the number of pulses sent, but they are all signalling the same thing: the period of that pure tone.

As the number of samples of the vibration increases, we would expect that the accuracy with which the period is provided to increase, and that our discrimination of pitch would improve. And like any poll which uses a sample to obtain a property of a population, once the sample is big enough, we do not expect the accuracy of the answer to increase significantly by measuring more. That is exactly what happens with our hearing (Tables 3 and 4). Our pitch discrimination of a pure tone, at all sound frequencies, increases with loudness up to a modest level of loudness. The sensation of loudness continues to increase from there on, but our pitch discrimination does not further improve. We cannot determine pitch more accurately than an adequate sample provides, and that limit is reached well before the loudness limit occurs. Would that the world realised one does not hear better by hearing louder than that limit!

As an aside, there are still people who believe (or want to believe) that pitch is indicated by where the basilar membrane is vibrating, rather than how it is. If that were so, the position would be most accurately indicated at very low levels of sound, and pitch discrimination would get worse with increased vibration, as the width of the region vibrating the hair cells increased.

Hz	31	62	125	250	500	1k	2k	4k	8k	(16k)
A	12	68	86	170	190	255	255	195	85	45
B	7	13	14	19	45	62	70	55	30	15
C	19	81	100	189	235	317	325	250	135	60

Table 4. The number of increments in the loudness of a pure tone that we can detect between the lowest level we can hear and the threshold of pain. A: Above the level where pitch discrimination does not improve. B: Over the range where pitch discrimination improves with loudness. C: Totals.

Measurements of loudness by black-box experiments show that we can only discriminate twenty distinct steps of loudness at the bottom of our hearing range, but over 300 steps in a 2kHz pure tone (Table 4). Our sensitivity to loudness change in the 2–4kHz range is surprising. It was selected for survival purposes, so there must be a reason. As a means of detecting approaching danger, 20 steps might seem quite sufficient. And in music, which is one of the few activities in which we overtly use comparative loudness today (Chapter 8), we only distinguish eight or nine dynamic levels at most. The scores of some modern composers have stipulated many more levels, but that doesn't mean that the players could produce them or the audience discern them, though I can believe that some listeners following the score imagined they did, whether it happened or not.

The only other way in which we are normally aware of our sensitivity to loudness change is in the unpleasant sensation of loudness vibrato. As we see in the next Section, we use our discrimination of 300 loudness steps for direction-finding. How the machinery can provide so many steps in some pure tones is a different question. At 2kHz, each hair cell is traversed by the travelling wave 2000 times a second. Let us suppose purely for illustration that at the threshold of hearing only 10 hair cells receive a firing level of vibration but at the upper limit of loudness 100 hair cells do. If on average a hair cell fires once every four vibrations, the range would be from 5000 to 50,000 pulses a second, ample to produce 300 steps of loudness. This is discussed a little further in Appendix 16 while Appendix 17 makes a novel suggestion of how we might remember music in a compressed form.

For the record, the sone scale which ostensibly measures loudness, requires one to measure the sizes of all the harmonics in a note or chord, apply a formula or scaled graph to each, and add the results together. One then has to do the same thing all over again for the next note or chord. With a tape recorder and a computer with a Fast Fourier Analyser program, to which one then applied the sone scale, it would take longer to assess how loud the first eight bars of a symphony were than to listen to them and obtain the loudness in a meaningful form.

10.3 The direction-finding system

As explained in Sections 1.3, 1.8 and 9.3, the ear-drum is vibrated by changes in air pressure, and pressure changes are the same regardless of the direction in which the sound passes the ear, but the delicate drum can be protected within a tube. In contrast, an animal ear primarily selected for direction-finding consists of minute bristles moved backwards and forwards in the direction the air vibrates. They must be exposed on the outside of the animal, where they are liable to damage. Insects use bristles which only resonate at the particular frequencies they want to hear, or even don't want to hear; mosquitoes can be discouraged by a battery-powered device which makes a noise like a dragonfly's wingbeat.

A pair of pressure-operated ears could locate sound direction roughly by loudness. If sound comes from one side, it will be louder on the side at which it arrives. If it comes from straight ahead, or from directly behind, it will be equally loud in both ears. But that does not tell whether the sound is in front or behind. We use reflection from our ear flaps. (We are, incidentally, almost entirely dependent on our ear flaps for discovering whether sound is coming from above us or below us.) If we think we have detected a very faint sound, we automatically turn the head through about 45 degrees to point one ear in the believed direction, our ear's most sensitive reception angle, but it needs the other ear to receive nothing: the rare commodity of silence. The advantage of rotating external ears is that they can be moved without revealing body movement. However, it might not be socially acceptable if we could rotate one ear towards a conversation at the far end of the dinner table.

If sound comes from one side, the far ear is usually said to be in the shadow of the head. What really happens is that the surface of the head absorbs much more of the high frequencies than the lower ones, so that when the sound reaches the other ear, the real difference is in the loudness of the higher frequencies. We are plagued by the monotonous thump of distant pop groups because low frequency sound is not absorbed by anything. That is why the very low frequency shock waves of earthquakes travel huge distances. So it makes sense that we have very good loudness discrimination in the 2–4kHz range. However, a few simple tests may question that. If we play the same music through headphones into both ears – it must be from a 'mono' source so that it is identical sound – and have one side at a slightly higher sound level than the other, we cannot detect any difference in loudness, let alone get any sense of direction from it. If we make one side so much louder that we can detect the difference, we still do not get any sense of direction; we just know it is louder on one side, and our reaction is that it is unsatisfactory. In fact, if we play a pure tone into one ear, and a different pure tone of similar loudness into the other ear, it is almost impossible to discern which ear is receiving which sound. Well, why should we? Ask a silly question; get a silly answer. Evolution didn't envisage our two ears receiving different pure tones. The diagram of the auditory pathway (Fig. 25B in Appendix 8) shows that there are cross-connections between the two sides from the very earliest processors in the system, and there are four further cross-connections between the two sides in the advanced system; one cannot think of any reason for that, other than that they bring the nerve pulses from the two sides together, to be compared, to determine sound direction.

If we can't tell which ear is receiving which of two pure tones, we have no hope of detecting small differences in loudness when sound arrives under normal conditions at one side of the head. Why does a small difference in loudness with music produce no sense of direction? Because in the real world, it would not be slightly louder on one side unless it also reached that side first. We don't determine direction in the cortex, we do it in sophisticated processors, so there is no reason why the cortex should be provided with small loudness differences. The processors' job is to initiate automatic actions by sending signals to appropriate organs. An enormous amount of automatic processing goes on to run our bodies, of which we are generally quite unaware.

The complicated and very clever part of direction-finding is determining the time difference between the sound arriving at the two ears. This is why we have the detailed coding of sounds which later made spoken language and music possible. The principle is simple enough. If sound comes directly towards the head, it arrives at the two ears at exactly the same time. As the head is rotated, it takes slightly longer to get to the ear further from the sound source, and the time difference increases until, when the sound is arriving from one side, it takes the maximum amount of time to get to the other ear. Rotate the head further and the time gap decreases, to be zero again when the sound source is directly behind the head. Neither time difference nor loudness indicate whether the source is directly behind or in front of us; we have to use the ear flaps to discover that. We can test it with another example of the susceptibility of hearing to suggestion. If we listen to 'stereo' sound in a room, that is to say, recorded sound which has a time difference between the versions from two speakers, naturally, or usually these days artificially contrived, the ear flaps indicate that it comes from the general direction of the speakers. Ear-insert phones remove the effect of the ear flaps. We 'hear' the orchestra in front of us, because we imagine that is where orchestras are. We can 'hear' the orchestra behind us if we want to. Try it; there is nothing in the sound which distinguishes the two.

It is not easy to decide the distance sound travels between arriving at one ear and getting to the other one when the head is side-on to the source; it is between 20 and 30cm, which means that the maximum difference in time is somewhere between 0.7 and 1.0msec. Our ability to allocate direction, suggests that the processors can use a time difference as small as 0.1msec, if not smaller. If one has the facilities, one can use a 'mono' source of music, and listen through headphones while introducing a time delay between the two sides. This produces a definite sensation of direction, moving from front or back to the side of the head as the delay is increased from zero up to 1msec delay. If we ask another silly question and continue to increase the time difference, the sensation is disconcerting. When the delay approaches 10msec, we may begin to hear sharp transients such as a side drum rim shot as two separate sounds. That parallels the experiments with loudness differences. We do not discern loudness difference until it is unreal, and we do not get a sensation of an actual time delay until it is unrealistically large, approaching the conditions when we might hear a real echo. The processors have to obtain the time differences; the cortex can't use time difference or loudness difference.

My colleague Dennis Unwin told me of a system used in receiving Morse in a crowded noisy radio wave band in which the sound to both ears was slightly delayed, but in one ear this delay was increased as the frequency increased, while in the other ear the delay was reduced as the frequency increased. This resulted in the wanted signal appearing to come from straight ahead, but the unwanted signals appearing to come from other directions. Many operators found that it was easier to listen to the wanted signal under these circumstances.

Our direction finder is more than sufficient for natural purposes, but probably not as good as we might think. A person blindfold and led in a circle at about three metres from a cello playing in a large hall, can tell which way the instrument is facing. At ten metres, the subject will often point to a place a couple of metres away from where the instrument is. In normal circumstances, we use other clues as well, such as where sound sources can be seen. Voices come from a moving mouth on television and usually on cinema screens, regardless of where the speakers are. Using ears alone we are probably accurate to about 10 degrees.

10.4 The timing system

The direction finder is located in the olive bodies (Appendix 8, Fig. 25B). The nerves are arranged in a highly organised fashion, with those coming from similar regions of the two basilar membranes, that is to say, transmitting the same ranges of frequency, in close association with one another. The cortex may not have a map of the cochlea, but the two medial olives in the brain do. They can compare pulse codes. We can only discuss this in principle but it is instructive. We cannot locate the source of a pure tone in a room. A constant-frequency sound produces pulses spaced with the periodicity of the sound, so the pulses coming from the two ears will be statistically identical, and if they are identical there is no way of discovering which of them arrived at one ear ahead of the other. We try to use loudness difference and our ear flaps, but the loudness depends on reflections from walls and ceilings as well as sound arriving directly from the source, and the apparent direction changes with every movement of our head. In 1951 I recorded a theme in pure tones for Anouilh's play *Point of Departure* produced by an undergraduate (now Sir Peter Hall); the sound was surreal and the source could not be located. The same idea was used years later for the *Dr Who* TV series.

The only way the system can tell that there is a time difference in arrival at the two ears is for it to 'identify' a bit of sound when it arrives at one ear, and to 'identify' that same bit of sound when it arrives at the other ear, and obtain the time difference. That is stating the obvious – except that the only way one can identify a bit of sound, is if it is different to the previous bit of sound, and different from the next bit of sound; in other words it is transient. We can only direction-find transients; all natural sound is transient. If there had been important constant-frequency sound in nature, perhaps we would have evolved bristles like insects. We rely very much on the starting transients to locate instruments, and the blindfold test described above was only fair if the cello was producing plenty of starting transients.

10.5 Sensing transients

The direction finder has to match the signals coming from the two ears, that is to say, match the periodicity in the pulse codes representing the frequencies, very precisely. So the sound must be coded sufficiently precisely to signal the frequency changes very accurately. The time it takes the signals to travel from the ears to the processors must be the same too. These processors were matching pulse codes a million years ago. Matching pulse codes is the basis of every way in which we have used them subsequently in hearing, so it should not be surprising if we use them that way in music too.

A noise consists of many changing frequencies. A typical picture of the basilar membrane at one instant when receiving a noise of several frequencies is a number of irregularly distributed lengths of resonant vibrations along it. But it is their changing that matters. An analogy which gives a simple picture of what happens can be derived from bar codes. Don't be worried; we are not going into how bar codes work. Think of several small vibrating lengths along the membrane as like the rows of lines of a bar code. Of course they are going to overlap, but transients are noises and noises are noisy! Take a pad of paper and print a bar code on the left hand side of the first page. On each succeeding page, print the same bar code, but starting each time one millimetre further in from the left for fifty pages. Flick the pages and the code moves across the page five centimetres. That is a way artists start to sketch a cartoon film. Take another pad, starting at the left side, but this time also gradually change the thickness of some of the lines and of the spacing between the lines, for fifty pages. Flick the pages, and the pattern changes in a complex way as it moves across. That represents a transient sound. We can print exactly the same thing but starting four centimetres in from the side of the page. When the pages are flicked, the pattern change is the same, but it happens in a different place. The pattern change of a transient is the same, whether it happens at a higher or a lower general pitch; words are like that.

We can alter the thickness of each line, and the spacing between each of the lines, in an immense number of ways, and when the pages are flicked, get a different pattern change. Each of those represents a different transient noise; those are patterns such as we recognise and store for words. But if we print the bar code in the same place without changing it, for page after page, that represents a pitched note with its harmonics, or a sustained chord. That is not much use to the direction-finder. If it was from an electronic keyboard with constant frequencies, it would produce a bar code literally staying in a fixed place. Once we start hearing it, it produces no new information. Heard one, heard 'em all, says the bored cortex. As for a pure tone, that is just a single fixed line of bar code. It is not surprising that in contrast to speech, noise or music, we cannot find any active area at all in a cortex when someone hears a pure tone; it conveys so little information that it does not stir the grey matter.

But the direction-finder is just an automatic device, though incredibly sophisticated. When a higher mammal hears a noise, the medial olive processors receive the pulses representing the changing pattern of the frequencies detected by one ear, and the almost identical changing pattern from the other ear. One of them is ahead of the other by a fraction of a millisecond. In some way, it obtains the time difference. Assessing the other way in which the signals are different, their loudness, appears to be the job of the other pair of processors, the lateral olives, and that is 'turned into direction' by co-ordinating the two in the next processors: the lateral lemnisci (Fig. 25B).

To achieve this, the ear's response to changing frequency must be very rapid. We now see a further advantage of the continuously changing resonance of the basilar membrane. When discussing minimal time to hear pitch (Section 9.11) I pointed out that any resonance needs a small number of stimulating vibrations to build up to full response. If, instead of a basilar membrane, we had, for example 1500 independent resonances, then as soon as the frequency changed, a new resonance would have to be built up. But let us say, for purposes of explanation, that a particular frequency has produced a resonant vibration on the basilar membrane distorting 100 hair cells. The frequency changes slightly. The vibration moves slightly along the membrane; ninety-five of those cells are distorted by the new frequency, five cease to be vibrated at one end, five new ones are now vibrated at the other end. The changed frequency does not have to build up a new resonance from scratch. Most of the cells simply change the rate at which they are fired by the slightly different frequency. A continuously changing frequency is signalled as continuously changing pulse codes.

A powerful computer can now synthesise speech noises quite realistically, no mean achievement, though it is only making copies of speech noises (and people can do that much more cheaply and easily). But I venture our direction-finding processors are a rather more remarkable device, as we shall now see.

10.6 Noise amalgamation

Once a processor has obtained the direction of a sound, we should not be surprised that it can automatically send instructions by pulses in nerves to various organs to point in that direction; that is a commonplace action in an animal. But what happens when two noises arrive at the same time from two different directions? We can certainly have a continuous awareness of two, one for example coming in at 9 o'clock and one at 2 o'clock. That emphasises that the direction is an integral part of each noise sensation. For one noise, the direction-finder has to match the changing patterns of the sets of pulse streams from both ears, to determine the time-difference between them; it will use all the available frequency changes in the sound to do so. Look at it the other way round. To the direction-finder 'a noise' is everything which has the same time-difference; that is the only thing a noise is to it. If it is also receiving a number of other changing pulse streams, and they all have in common another time-difference, that is a different noise. It adds any help it can obtain from their respective loudnesses, but time difference is the only thing which will sort a collection of simultaneous pulses streams into packages which come from different directions.

It is everyone's experience that the two noises (as pulse patterns) are sent to the cortex packaged: amalgamated as two distinct sensations with their directionality. How, we don't know. We have reason to believe that in other mammals who cannot identify more than half-a-dozen different noises, the thing their cortex does receive is the directions that noises come from. And if a processor automatically packages all the frequencies in a single noise into a unitary sensation, it will certainly package all the *harmonics* of artefact noises which are actually music notes as a single noise. In other words, amalgamating harmonics into a single sensation is a basic function of the natural system. It will use whatever little transience the starting transients and timbre that sounds contain to package them, and if it can't do that, we can't obtain their separate directions. It is another reason why real instrumental sound is more satisfactory than constant frequency electronic sound. And loudness vibrato does not help direction, whereas pitch vibrato does.

In what form are the messages which indicate sound direction? There is one tiny straw at which to clutch, described in Appendix 18. But there is a remarkable example of how our automatic processors can sort, match and amalgamate pulse codes, for which I am indebted to my colleague Robin Walker. During the Cold War, the BBC's European Service radio transmissions to Eastern Europe on two or three different radio frequencies were jammed by music or noise broadcast from behind the Iron Curtain on each of the same radio frequencies, to prevent anyone listening to the BBC programme. But it was different music or noise from each of the jamming stations. It was discovered that if two radios were used, one tuned to one radio frequency and one to another, and connected to headphones so that each ear heard only one radio, the ears received the confused noises but the programme could be heard in one's cortex. The processors recognised the pulse codes from the two ears which corresponded, selected them, and sent the amalgamated signal to the cortex. And if you don't accept automatic processing after that, I won't expect you to believe anything else in this book.

10.7 The general characteristics of hearing

The characteristics of hearing such as the inter-relationship of frequency, loudness and discrimination, are usually shown in scientific texts in a series of complex graphs. I put one of them in *The Violin Explained* (1997), because I went into the detail of the sound of the violin family. I think we can see all we need for this over-view in simple tables. The values in Tables 3 and 4, gleaned from various sources, are presented in very round figures in a form which has some musical meaning; for example, if we can only discern about half-a-dozen different pure tone pitches in a semitone around middle C, what is the point in giving values in more precise physical measurements?

Amongst the things we can see from the tables, is a very real difference between frequency and pitch. We discriminate the pitch sensation in the musical terms of how many pitches can we separate in a semitone. It is very poor at the bottom end of the hearing range and improves continuously right up to two octaves above treble staff. The bottom line of Table 3 gives exactly the same data for moderate level pure tones, but expressed as what difference in frequency, that is, what difference in the rates of vibration of two pure tones we can detect. They are remarkably similar up to about 2kHz. An interval in music is not a difference in frequency (though I'm afraid that is how it is described in several respectable music reference books); it is a ratio. But the direction-finder does operate entirely on the difference in frequency, regardless of what the frequency is. That is the only thing it can use, and so that is the property it selected in our ears.

Why does our frequency discrimination get poorer at around 2kHz? The answer is apparently very simple. The direction-finders in the medial olives only compare the pulses which come from low frequencies up to about 2kHz. If we look similarly at loudness, Table 4 gives the number of steps of loudness we can discriminate in pure tones below and above the level where our pitch discrimination reaches its limit. There is no useful loudness difference between the two ears with low frequencies. The head absorbs the high frequencies; that is where loudness difference between the two sides of the head is useful to the direction-finder, and that is why we have our best loudness discrimination with high frequencies. It would have been the value of high frequencies in direction-finding which caused natural selection to extend our hearing range to 4kHz and higher.

Thus it appears that *the direction-finder determined all the properties of our hearing*. It required frequencies to be separated and coded, but only as much as it needed them; accidentally that gave us pitch, but with very different accuracy at different frequencies. The coding enabled us to select pitches which relate. It gave us the amalgamation we know as tone, and an ability to improve the selection of related pitches by it. It gave us a loudness characteristic which ascribes additional importance to the sounds in the frequency range where loudness is most important to direction. And it attaches the greatest importance to obtaining transience and timbre, and the exact timing of the receipt of sounds, because that is what it uses. And when we invented music, everything was in place to use the sounds. Everyone gets the sensations automatically, and as I have said repeatedly, a person need not perceive anything about music to enjoy the sensations. What I have not said before is that applying the cortex to those sensations produces very mixed rewards. We may enjoy some music more but other music less, and we may certainly enjoy some performances of it less. Perhaps the noble savage is still happier than the servant of a highly attentive cortex.

Our sensations are a distorted version of sound. All we can do is use them as they are. Fortunately, we don't hear music as we hear pure tones in laboratory test conditions (nor, recording technicians please note, do we want to). Because if we did, no player could produce pitches sufficiently precisely. It is very doubtful if anyone can judge the pitch of an orthodox instrument's intervals to better than a tenth of a semitone. We may do a little better with simultaneous simple intervals if the harmonicity is excellent. But all this only applies to using instruments in the way for which makers painstakingly developed them over three or four centuries. If composers try to push the range higher than that for which makers designed them, players have increasing difficulty in controlling the pitches and tone, and listeners have decreasing difficulty in judging them! In every walk of life, tools always work well if they are used as their makers intended.

10.8 Space sound

Lastly, we come to 'stereo' recording and space sound. Hearing evolved in the open air. In natural open space, sound normally has unconfused direction. In any enclosure we hear sound direct from source and with reflections from walls and ceilings. If the purpose of a stereo-recording was realism, which is often claimed, there would be only one way to record it, and that is to use nothing more than two pressure-operated microphones where the ear-drums are, in a dummy head, in a position where a member of the audience would be at a concert or recital. If we then listened through headphones, we would, in so far as it is possible, obtain as realistic a performance as if we were at the concert. That would preserve all of what little time and loudness differences there are in a concert hall, including all the reflections. It would not only be realistic: it would also be very inexpensive both when recording it and when listening, and environmentally friendly. It isn't done that way. Multiple microphones are used, often in places one would never listen to anything. There are recordings which make one feel one is sitting on the platform between the instruments, or with one's head inside a grand piano. The balance between instruments is decided by the technicians' tastes, which amongst other things can result in solo instruments being too loud, and if high frequencies are too strong, the tone is altered. The sound is then reduced to two channels, with artificial time differences put in. It is often not realistic. If such expensive and unnecessary technology must be used, one would at least wish that a musician make all the judgements. Scholes (1967) records that after an argument in 1925 at the BBC, the Corporation's Director of Music, the Professors of the Royal Academy and Royal College of Music, and Scholes listened by radio to three orchestral works played twice, one performance of each being controlled by either an engineer or a skilled musician, but they did not know which was which. They were unanimously in favour of the musician's judgement; the lesson seems sometimes to have been forgotten.

Most modern recordings have unnatural 'clarity', even sometimes including those unavoidable noises inherent in playing instruments, such as the plops of woodwind pads, and the sliding of fingers over strings, which one does not hear at any civilised distance from instruments. Until 1950, composers could assume that their music would be heard with appropriate reverberation from their surroundings. The simple time-honoured way of testing an auditorium is to clap one's hands and hear how long it takes for the reflections to die away. And although it may affect different frequencies differently, in general terms that is the extent to which every sound is extended. Gabrielli composed some fine antiphonal music for St Mark's Venice, on the assumption that there would be reflections lasting ten or more seconds, and it needs it.

Sound extension can be examined using an electronic keyboard with facilities usually labelled 'reverberation' and 'chorus'; use imitative piano sound, because it is entirely transient. (If possible suppress the keyboard's speakers, by using the headphone socket to feed a good amplifier and speakers. The signals keyboards generate are often respectable, but their amplifiers and speakers often are not.) Both effects are produced by adding to the original sound identical sound somewhat reduced in amplitude with time delay. Up to five or so milliseconds between the original, and the same sound delayed, produces 'richness'; if it is extended much beyond that, the impression is of a piano playing in a large hall with reflective walls. The perception problem is that if we start with unadulterated sound and then extend it, we know it becomes 'fuller' or 'richer'; if we did not know the unadulterated sound, we might accept the extended sound as a normally rich piano. We are not deceived by the marble hall effect, because we can see that the piano is in a small well-upholstered sitting room, but we could not deduce that if we were blindfold in a strange place.

The standard recommended design values for auditoria are about one-second reverberation time for speech and up to two seconds for music. We almost always hear live music with more than that amount of continuing reverberation, and in relation to the rate at which notes succeed each other in most music, a second is a long time. In general, we don't like music in a 'dead' hall. Open-air performances of serious music (which are also occasionally broadcast) always sound unattractive. Players loathe doing it for several very good reasons, but the sound lacks the reverberation extension. And an instrument played in an anechoic room with completely sound-absorbing surfaces is almost unrecognisable. Music is artificial, but so also are the circumstances in which most of it is heard. Have we become conditioned to wanting the continuing sound? So-called 'surround sound' is produced simply by having an additional speaker which adds sound delayed by some milliseconds to that coming from the main speakers. It is created commercially using digital sound and a silicon chip producing the delay with the splendid name of a bucket brigade. But one can try 'surround sound' very easily. Sound travels at about one foot (30cm) per millisecond. Adding an ordinary 'mono' radio at least ten feet away from the speakers of one's normal stereo system and from oneself, produces it. In my experience most people find the novelty wears off very quickly and don't like the gimmick, any more than they like their piano in a marble hall. But it does kill the excessive synthetic clarity of some of today's artificial stereo recordings.

10.9 Conclusions

There are many impressive things about our hearing system, not least the extreme accuracy with which it preserves the timing of transients. That sensitivity to time is a feature of the one basic aspect of music we have not examined: rhythm, which is why that is in the chapter which follows. The purposes for which our hearing system was naturally selected, it achieves almost miraculously. But if we use any machine for things for which it was not designed, we must expect to get some odd results on occasions – such as music.

This has been my much simplified account of hearing, though I fear it will still appear complicated to some readers. If it does, I can only say, with no disrespect to the specialists who are researching an exceptionally difficult subject, that their knowledge is now a huge morass of facts in complex technical language in which it is often very difficult to see the wood for trees. All I have tried to do is to create a model from the mass of facts, which appears to make sense for the basis of western music. Of course it does not explain everything, and it is easy to criticise a model because it does not explain some particular property about music. May I hope that anyone who does will also be constructive and provide a model of the mechanism which answers the criticism. Music is only a small specialised aspect of the larger subject; it does not seem to be of particular concern to many of the scientists involved, which is reasonable, and just as well to judge by one or two who have presumed to refer to it. One does need to know just as much about music as about hearing before trying to put the two together.

I hope I have provided some kind of answers to the questions listed in Chapter 9. There is a basis for being able to sense pitches in the simple ratios of the pentatonic and heptatonic scales, why simultaneous seconds are not comfortable, and simultaneous semitones are even less so. We have some explanation of how the sensation of sounds with harmonics in simple ratios have the characteristic of the pitch-frequency and the amalgamated sensation of tone. The seventh harmonic, which we probably never obtain distinctly, disappears in that package like all the others. Harmonics higher than the seventh make little contribution except in causing our ears to generate buzz and hiss, which is actually a characteristic of all complex sounds. We can see why minute continuous variation in harmonics produces timbre, and why vibrato may add to that until it is observable.

The packaging can create a missing fundamental, and is essential in augmenting the pitch characteristic for low-pitched instruments, and in instruments with weak resonance in the low part of their ranges. We have a basis for loudness, and for the ease with which we hear high-pitched instruments or those producing harmonics in the most sensitive part of our hearing range. The actual frequency range we use in music, however, does appear to be based on physics – the instruments themselves cannot be made small enough to produce very high-pitched sounds accurately. We have a basis for stereo-sound and all the other tricks played on our automatic processing system. That is where the cortex comes in. We can readily believe things about musical sound which we cannot perceive, and we can fail to observe things which are observable, if we don't want to hear them.

Two things emerge from the discussion. Sensation is instantaneous, but perception has duration. We cannot observe anything about sound or make use of it unless we use memory to observe it over a period of time. And the more we can attribute to mechanism, the more we can expect those who speculate on what happens in the cortex to work from a basis of what the mechanism presents as sensation. There does appear to be a physiological basis for orthodox tonal music. The way in which sounds are coded offers a basis for pattern matching between the components of simple intervals and chords. It is difficult to avoid the conclusion that that is why those sounds were originally selected. It would follow that if the sounds are so related, it is relatively straightforward to obtain what the music is trying to communicate. If the sounds are not related in that way, it is infinitely more difficult for anyone who does try, to discover by listening how they are related, and on the assumption that they are, what the music is trying to communicate. That does not mean that we should all like only tonal music; it probably does mean that most people always will.

11. Time and rhythm

1.1 Introduction

Rhythm is often considered the most difficult feature of music to understand. I have left it until after considering the hearing mechanism for at least we now know why it is so time-sensitive to the arrival of transients. There is, however, a more fundamental reason for considering it as a quite separate phenomenon; it is not a sensation like sound produces. There are also terminology difficulties. 'Rhythm' is often used unselectively for almost anything concerned with the time characteristic in which music is played and heard, and sometimes interchangeably with 'the time'. The word 'beat' is used, and in reference books one finds 'when there are four beats in a bar', but no attempt anywhere to describe or define what a beat is. What a conductor provides is also called 'the beat'. The term sometimes used by players is 'the pulse' which has the virtue that it is only used in one sense, as a feature of playing.

The automatic machinery of people with normal hearing offers the same signals to everyone's auditory cortex. From then on, how an individual responds to the sensations produced depends on that individual's cortex. Amongst their responses they may react in a regular way as a result of the time of arrival of transients. We will call that a *rhythmic response*. If we assume that it is entirely a cortical matter, there is no difficulty in accepting that it varies enormously between individuals, from zero to the almost compulsive. It is a consequence of hearing music and, in a subtle way, of having heard music, as we shall see.

Sound that conveys such a regular time pattern has to be produced, and, mechanical and electronic generators aside, it is produced by people. They require a time-base upon which to produce the transients. If the player has an awareness of that time-base we can call that *the pulse*. Both the pulse and the rhythmic response are mental activities not sensations. They are normally characterised by having a very simple distribution in time but they are not necessarily synchronous with transients of the music being received. Neither are usually at the fully conscious level. The pulse is, and the rhythmic response may be, directly associated with physical activity. This is quite different to the sensations of pitch, loudness and direction.

11.2 The origin of time-patterned sound

A transient occurs when any resonant object is struck, and we made noises when making and using tools. The idea of making noises in repeated *patterns* in time has to come from somewhere. I dismissed in Chapter 1 the suggestion that it may have come from animals running in; what purpose would be served by imitating that? An individual may walk in a regular pattern over a relatively level natural terrain, making little sound in bare feet or simple footwear, but small groups don't naturally walk in step even today, and anyone who has watched (or been one of) raw recruits knows that marching does not happen automatically; someone has to yell the timing transients. I believe the development of making patterned percussive sound for its own sake would have been associated with communal activity, and in the broad sense with pleasurable activity which required a reasonably settled developing society. I find it persuasive that regular noises were produced by ornaments attached to the arms and legs of people carrying out primitive repetitive group movements. The movements must have been copied visually. It isn't easy even today to tell people what to do; it is usually demonstrated. We met that principle in Chapter 5 when discussing 'teaching' children to speak or to play instruments.

Sound produced by noise-making ornaments in roughly imitated movements is self-reinforcing; the greater the synchronisation of movement, the more the sound is also synchronised. A typical modern example of this is an audience clapping. The individuals clap randomly. If coincidentally or purposely a small group clap regularly synchronously, the sound may pull the mass into regular clapping or foot stamping. They imitate the rate. The jump from hearing the repetitive sound from 'dancing' to making sounds by hitting a suitable object at a similar rate is small. The sound produced is more defined, and if an individual can store the sounds in short-term memory, the beating may be spaced with corresponding accuracy. Each repetition reinforces the system. The determination of the rate begins to pass from the dancers to the percussionist.

Each repeated movement of 'dancing' takes a finite time. The actions occur at a regular rate but a transient produced at that rate could mark any point in the repeated movement. If the movements are to be synchronised the transient has to mark a particular point in the actions. This establishes the three basic elements of synchronising communal group movement, of dance in its most basic form. Each individual has to know in advance what to do, what movements to carry out. They have to know the rate at which the actions should follow one another and repeat. They must get a signal to indicate when a particular feature of the repetition should be initiated to synchronise their movements. Suitably interpreted, it is also the basis of the ensemble performance of music.

Since using related pitches was certainly discovered independently several times, there is every probability that making repeated sound patterns was too. Simple experiments with different noise-making devices would lead to the discovery that maintaining regular percussive sounds with long intervals between them is not easy (you can try simple experiments *without auralising music while you do* if you want to test this). It is easier to do so at rates of a few sounds per second but it does not provide longer-period synchronising signals. Simple patterned sound produced either by varying the loudness of equally spaced noises, or by producing the same sequence of different kinds of noise equally spaced in time, when regularly repeated, provide a synchronisation signal, and it appears easier to store a pattern and repeat it than to maintain a simple regularly spaced single sound. Nor should we think it was necessarily done only with pitchless percussion instruments. A solo player can produce effective repeating simple patterns with instruments using very recognisable but unrelated pitches if they have strong starting transients such as from tensioned bamboo strips, nail pianos, xylophone bars and indeed panpipes. We should not think that all pitch-based and pattern-based music had separate origins. Whatever happened, some discoveries then developed into music predominantly of complex time-patterns and some into forms in which the main interest was in pitches.

11.3 Time in pitched music

When selected pitches on a simple wind instrument progressed from extempore sounds to making them in a similar sequence from performance to performance it was stored in the creator's cortex and it was transferred aurally to others. The sequence of pitches is one part of its identity and it might appear that the durations of the pitches are the other part. But within certain limits, those who can store and recover a tune can do so in *relative time* for it may not be recalled at the same tempo as previously heard, and, in speeding it up or slowing it down, people maintain the proportionality of the occurrences of the pitches in time.

We store a melodic line as relative pitches, and their distribution as relative time. We have a basis for relative pitches, because pulse trains represent pitches, and pitch relationships can only be recognised as relative pulse rates, as ratios. Superficially, storing time characteristics on a *relative* scale appears to be a very complicated achievement, carrying out a continuous proportional adjustment of durations while recovering a tune in auralising, vocalising or playing it. I don't put anything beyond a cortex if it puts its mind to it (sorry!), but anything which many cortices seem able to do without thinking (even bigger apology!) usually has a relatively simple basis. And if almost anyone who can recall and sing a tune from memory can achieve relative adjustment of the time spacing, then repeating a percussion pattern is trivial. We don't assess the durations of the pitches, or the time intervals between transients in absolute terms, because we can't; we do not have any absolute ability to measure time. We locate the occurrence of the pitches, relative to a regular *time framework*. That framework is part of the identity of the melody. It has to be of a relatively simple repeating form, such that we can speed it up or slow it down, and that enables us to locate the pitches in it, at the rate we set.

11.4 Time and notation

Despite the role of the time factor in the aural transfer of song and dance over millennia, it apparently took a very long time for the integral nature of a time framework to be recognised in notation, though since notation was largely developed in ecclesiastical centres which were mainly interested in chanting and plainsong, that may not be so surprising. But when it was recognised by the invention of the *bar and bar line* (*measure and bar* in American terminology), it was adopted with remarkable speed, virtually within a hundred years through-out Europe. The bar, and grouping notes and rests into the sub-bar units of the time framework, enabled players to *see* the relationship between the notation and the time-frame. What they also needed was the time-frame itself, and that of course is what they call the *pulse*.

11.5 Obtaining the pulse

So far as the individual player is concerned, with music which is in a repeating time framework the music provides the pulse and the pulse determines when the components of the music are produced. It follows that during normal playing, the player must have obtained the pulse from the previously heard music, and playing continues to reinforce it. It can be described as feed-forward rather than feed-back. There are several matters that flow from that generalisation, not least that the rate of the pulse has to come from somewhere to start with.

The experienced soloist will inevitably determine the rate before starting to play; the rate will begin as some form of activity in the conscious, continue for the first bar or so heard, and then run in the 'background' from then on. Playing in ensemble is no different in principle except that the sounds which have determined the pulse and reinforce it are provided by everything one hears and not by just what one plays oneself. Often in cathedrals, and especially in the open air, hearing others can be very difficult. In chamber music one could never count bars of rests unless one could continue to obtain the pulse from the sounds, often bringing it to the surface and counting to the pattern. But if the pulse is reinforced by 'the previous bar', it has to come from somewhere when there isn't a previous bar – at the beginning of the piece of music. There are all kinds of ways in which small ensembles arrive at the rate, by discussion, counting (very audibly with Dixieland groups), foot tapping, nodding and so on, before starting to play. Even a well-rehearsed string quartet may have covert signals, and will watch the leader's bow like hawks for the all important first bar. For larger ensembles, there is a conductor. That deserves a separate section.

11.6 The conductor

There is a widespread belief amongst those who have no experience of orchestral playing that the conductor provides 'the beat' like a visual metronome marking the rate of the time framework. Some do, and in some circumstances such as a change of tempo or time framework, they should. But of course they cannot provide the pulse, because that is an internal function in the player; hence if a conductor does display the time-frame we can call that the beat without confusion. However, the first advice one gives an embryo conductor when faced with an experienced orchestra is that one does not conduct to points. All the movements of the stick or hand should be smooth when the beat is steady, except for the 'click' as Adrian Boult (1968) so aptly described it in his excellent little book. This is a tiny flick of the stick at the top of the up-beat, a rare case of using an aural analogy to describe a visual one. And the click is the synchronising signal described in Section 11.2 above. In conventional music everything else should be provided by the players' pulse. Where the time-framework is continually changing, for example in some passages of the *Rite of Spring*, the clicks and beats are needed. And if the music doesn't provide any sound pattern which sets or maintains a pulse, then the players are entirely dependent on the conductor, and if they have to read complex notation at the same time, it puts an incredible strain on them. This point is amplified in the following section.

It is an under-statement to say that conductors have the responsibility of ensuring that the orchestra all start together, because their real job is to provide the time-frame rate; indeed one of their prerogatives is to decide it. And there is an extraordinary idea that it can be indicated by the 'speed of the upstroke of the baton into the first beat'. Unfortunately many conductors not only believe it but think it unprofessional not to do so. I know many experienced professional players who have begged opinionated young conductors to give one-for-nothing for some pieces – and been refused. That is one of the real misnomers in musical language, because it is one for the vital rate of the time-frame, and the rate cannot be communicated in any other way.

What else a conductor does is as varied as the number of conductors, and if you haven't played under two dozen of them, television provides ample opportunity to see the variety from the minimalist to the extravagant, whether they are ahead of, with or behind the orchestra, and whether you could determine a rate from a single upbeat. Of course the conductor's major contribution is made before the concert. And there is an apocryphal remark made about that by a leader after a concert. 'The rehearsals are for you, maestro. The performance is for us.' I leave the circumstances to the reader's imagination.

11.7 Time and the beginner

Discovering how to play in time from notation is exceptionally difficult. The natural expectation is that symbols which represent durations would be spaced according to their durations, which is how things are shown in almost all other visual representations of time. Notation uses spacing which reduces the allocation for long notes and expands the space for short notes. Having conveyed the principle that the symbols, not the spacing, represent durations, children are left to divine that they represent comparative durations.

Because of the complex process of reading notation and establishing a visual link to a physical operation, hesitation and irregular timing are bound to occur, and they do not provide any pulse. If the timing of a bar gives trouble, get the pulse from a couple of previous bars and then play it. And there is a far more important reason for discouraging foot-tapping than that it does not look nice or it might make a noise. Never acquire the habit of tapping the time while practising a monophonic instrument on your own; feet are tapped as a result of hearing music and not as a determinant of time. There was an oboist in one amateur orchestra who was known as the man who played in time with his foot and not in time with the band, because he always practised that way. For a young wind or string player, someone who can provide a pulse from a simple keyboard part is invaluable, and pieces with keyboard accompaniment for learners should always have a short introduction which establishes the pulse.

When children start playing in their orchestras, they have the further complication of being told to watch the conductor. This is distinctly unfair because when we become experienced orchestral players, we know how little we do, and very often, how little time we have to do it, when we are reading complex notation. The junior school orchestra is ragged because most of the players are trying to get the time from the conductor. There is a gradual transition as orchestras and players improve until most of the timing is done by hearing, though one cannot explain it to young players. The ultimate case of timing by hearing can occur with a small orchestra in a band pit in the theatre where a double bass playing pizzicato can control the tempo more effectively than any conductor; ask any experienced player if you don't believe it.

11.8 Variations on a pulse

Orchestral music has to be played with the exact timing proposed by the notation, otherwise it would be ragged. That doesn't prevent it from producing a rhythmic experience in susceptible recipients. Marches and waltzes can do that by the nature of the time-patterns conveyed by the sounds. The common and originally deprecating term for all music so played is *square*, but it is a valuable term. When jazz arrived, it became very obvious that some of the tunes were not played square, that is to say, the starting transients did not exactly correspond with the very strong time-framework that the music conveyed; we can't say whether the tunes corresponded with notation, for many of the tunes were not written down until after they were well known, and the composers often could not have done so anyway. Playing in this way is described as playing with *lift*; the meaning is simply that to some recipients it enhanced their rhythmic experience. The term isn't restricted to jazz because it is just as applicable, for example, to Shetland fiddling for dancing (Cooke, 1986). Unlike square, which has only one meaning, not all timing variations produce lift, or those which can be introduced into the music by the programmes used for setting music on a computer would – and they don't. In fact, for all sorts of technical reasons, computers do not play back polyphonic music precisely square either and one can hear that it isn't. But they are invaluable for 'proof-reading' scores by ear instead of eye.

Playing with flexible timing, or 'with rhythm' as it is often called, is normally a prerogative of a soloist, and one may interpret 'with rhythm' literally, in that in jazz and similar improvised music the soloist invariably plays with other instrumentalists who are providing the listener with a strong regular version of the time framework as a basis against which the soloist's sounds are heard. The typical listener concentrates attention on the soloist. The 'break' when the soloist is unaccompanied ceases to be effective when the listener can no longer maintain the time-frame. Up to a point one can compare the reaction of the listener to the pulse essential to the player, but it is frequently associated with foot-tapping and similar regular movements as a *result* of hearing the music. That is entirely different from using the foot to determine the pulse.

11.9 You can't write it down

Solo playing with very slight variation from the square timing which ordinary notation indicates, is a feature of an enormous amount of worthwhile performance of all kinds, and it seems likely that J. S. Bach recognised this or he would not have put a time signature of $\frac{18}{16}$ on the twenth-eighth Goldberg variation, which must mean that he did not want it to convey any pulse or time variation. But whatever the kind of music, one cannot indicate flexible time by notation. Couperin invented two signs which requested 'play ahead' and 'play behind' the notation value, and one may doubt that that achieved any greater success than any invented hieroglyphics which those of us who have wanted our music to be played with lift have used. Lift is an inherent ability, and the composer is entirely at the mercy of the performer. Cooke wondered whether Shetland fiddlers copied it, for their music is acquired entirely aurally, but the elders can recognise recordings of individuals, and to say that people copied a style does not really mean anything. It cannot be taught. Indeed when I demonstrated how I wanted a piece in one of my works played, to a technically superb undergraduate pianist (now a distinguished cathedral organist) I was reprimanded with 'That is taking unjustified liberties with the written music!' I wonder how many hemidemisemiquaver triplets would be needed if I had tried to score it for him as I wanted it played.

A different side of the picture is provided by the sophisticated transient patterns of latin-american music, which nevertheless communicate a very strong rhythmic experience to some listeners. They do have to be acquired aurally and can't be written down. Milhaud, Revueltas and many others have transcribed them – as square versions, and if they are played as they are written they sound square too. The only thing one can ever do is write down a square version and hope for the right kind of performer. Latin-american patterns are extremely effective if suitably played on a harpsichord, so that their effect has nothing to do with the emphasis of notes. The harpsichord is a valuable tool for distinguishing what is conveyed by timing and what by emphasis.

11.10 Metronomes and click tracks

Musicians hate playing to a metronome click; metronomes are useful for providing a rate before playing and for putting a tentative rate on a score, though a broad indication is often best. Performers are musicians, and the more one treats them as automatons the less one can expect them to contribute to the music. The metronome is a good example of this. The player must already have the pulse internally. And many recent recordings of pop music demonstrate how music is killed by a metronome for they are as square as a draftsman's T. For the convenience of recording engineers, each player has to record their part on a separate track while listening to a click track – a metronome – and the clicks are then used to synchronise the tracks while the technicians adjust them to their taste and mix them. I know talented young musicians who can't do it; we can understand why. Nothing compares with a recording of a live performance in which the players provide each other with the time-framework.

In this context, although investigators have tried, there is no way in which one can measure flexible time in, for example, a performance of classical music by a pianist. Using MIDI from an electronic piano one can record on a computer when notes are pressed with an accuracy of perhaps 2msec. The simple way might seem to be to include a marker from a metronome. But as we have just said, if you want to kill a musical performance, give the player a click track! A flashing light is just as destructive. And if there is no reference marker, where is the pulse against which the music is played? It is in the player's head. One can demonstrate simple things, such as whether a pianist speeds up on an ascending scale and slows down with a descending one, but in an effective live performance even bars may not occupy exactly the same length of time.

11.11 What is rhythm?

That has been a brief discussion of time phenomena in music, though most of it will be things that many musicians know instinctively. And that is the problem. I haven't defined rhythm and if you ask me what it is I can only echo the famous reply of Duke Ellington: 'If you ask, you ain't got it.' Unlike anything else in music, some have it, some haven't, it can't be taught and it can't be acquired. And some people can't tell whether music is played with it or not. I was lucky. As far as I know, I was born with it.

12. Conclusions

12.1 We all hear the same thing

A basic premise of this study is that the hearing system, up to the point where it delivers signals to the auditory cortex, is an automatic mechanism, and therefore all people with normal hearing receive the same sensations from the same sounds. It would appear to be amply supported by psychophysical testing with pure tones and by medical audiometry. It is also tacitly assumed in any discussion between two people who have heard the same piece of music; they may have observed different things and have reacted very differently, because those are cortical processes, but people do not commonly say 'Well, I got a different sensation.' If they did there would be no basis for criticism or review, or indeed for music itself. On such grounds it is legitimate to call the system up to the point of delivery a 'machine'. But why our hearing behaves in this way has nothing to do with music or speech. We discovered there is reason towards the end of Chapter 10. Our sense of the direction from which sounds appear to arrive is absolutely dependent upon the faithfulness with which the detail of transients is supplied to our direction-finders. Our survival depended on it; we still have that ability in every respect. Natural selection determined how the ears should work, and it obtained the best possible machinery from the potential of the living material to achieve it. And that is why we now hear what we do hear in the form that we get the sensations. Those are also the grounds for suggesting that our hearing system has not changed for a hundred thousand years or more, and, since higher mammals who have excellent directional hearing have very similar anatomical structures, the reason for saying that it may not have changed since the emergence of mankind, a million or more years ago.

In addition to the nerves which carry the signals upwards from the ear to the cortex, there are nerves which run in the opposite direction as there are in most of our body's systems. Nerves are required for many things, not least to help maintain body functions of which we are quite unaware. There are return nerves which tense the tiny muscles of the middle-ear bones in the presence of very loud sounds, just as there are in our optical system to contract the iris in bright light (in technical terms this is feedback producing volume compression); neither affect the sensation of loudness or brightness because the systems take account of the control, and they are automatic too. There are also vestigial materials no longer needed but not eliminated because they are not disadvantageous. Perhaps those who can wiggle their ears have vestiges of the system vital to animals who rotate them. But the significance of this paragraph in our present study is that we can't alter the sensations we get by taking thought. Unfortunately we don't have anything like eyelids to turn our hearing system off; that is our present penalty for its past importance.

12.2 The origins of music

No one knows how music originated, but in order to discuss the possibilities a little chronology is essential. In 20,000 BP (before the present), the last Ice Age was in decline; there were still woolly rhino and mammoth in France. We were, as our ancestors had been for a hundred thousand years before, nomadic hunters who travelled in tiny family groups, following and living off the wild herds of animals as they migrated north, and attacked both by large carnivores who also followed the herds, and by other humans. We carried little more than a few hunting weapons and small children, and sheltered in caves. Like most other animals, we were naturally selected not to make a noise. Few people today have had the experience of being soldiers, but films in which treading on a twig makes the difference between life and death are common. Hunting is now a controversial activity but every bird watcher knows the virtue of silence. The first settled communities such as Jericho and Jarmo in the Middle East, and Tehuacan in Mexico, date from about 15,000 BP. Settlements in the Far East, the Indus Valley and coastal Peru appear to be a little later. They began as groups of just three or four family units, and with life expectancy and infant mortality levels of that time, perhaps two dozen individuals, who spent their entire lives tending animals, gathering grain and fruits and fending off animal and human enemies.

My deduction is that at some point after that someone accidentally discovered steady-pitched sound from a natural tube blown as a panpipe or flute. When, no one knows. But if seven-holed flutes were used in China in 9000 BP with a scale approximating to simple pitch-frequency ratios, and one follows the argument in Chapters 2 and 3, the time-consuming experiments which led to pentatonic instruments would have been started a few thousand years before then. They could have been carried out by just one person in a tiny community. The instruments made little noise. Such experiments gave rise to the same pentatonic scale on very different kinds of artefact in half-a-dozen completely isolated locations.

Other than the legendary gift of music by the gods which occurs in almost every folklore, the one alternative possibility which needs consideration is that pitched music originated with the human voice. We pointed out in Chapter 5 that apart from a few involuntary noises, our use of the voice is almost entirely imitative. Someone in each tiny nomadic band 'invented words' but there would have been very few of them until settled communities started. Suppose there were steady-pitched sounds from natural animal noises and that a wolf or coyote did make one. For what reason would anyone try to imitate them? We herded goats; the last thing we would do is attract predators. It is more realistic to imagine we whistled to attract and snare birds. If one wants to argue that steady-pitched sound originated from people rather than artefacts, whistling is a more promising starting point. The first question is why anyone produced a steady-pitched vocal sound, such that it might form the foundation of a pitched music system. It would have been accidental, for in learning to speak we try repeatedly until we manage a word. It would have been ephemeral. Why should it have been remembered and remarkable? The magic of the panpipe tube was not only that an inanimate object made a sound, and a sound like nothing which had been heard before, but that it made the same steady-pitched sound every time – a reference point which facilitated future experiment and discovery.

One steady pitch is not music. Two steady pitches is where it begins. If a voice could make one, it could obviously make another. Indeed if we follow the argument of Chapters 3 and 6 that in the comparison of pitches, hearing is concerned only with the repetition rate of the pitch-frequencies, and the cortex with the consequent patterns of the pulse streams generated, then voices could have given rise to simply related pitches, just as we can argue that way about artefact sounds. To have had any continuing existence it would all have had to be remembered, whereas pitches from pipes are embedded in the hardware. However, on the evidence of Chapter 4, I think we can abandon theories such as that a multi-pitch system arose from the harmonics of the voice, just as we should about the harmonics of artefact sounds. Although the vowel character-istic of the voice is obtained by providing the mouth cavity as a resonator for one (or more sometimes) of the harmonics of the vocal chord vibration, in no way can we use that as a source of a second pitch. The vowel sensation is *tone*, the sensation of the amalgamated harmonics. Helmholtz showed that singers can adjust the formant (another term for resonance) of the mouth to match a harmonic, but it can't be done the other way round – one can't make a harmonic match the mouth resonance; one can only emphasise a harmonic already present, and the mouth resonance is comparatively broad. Since voices vary enormously in their harmonicity (Chapter 5) it would only be a voice with near ideal harmonicity which would produce a vowel formant making a simple interval. There is no such thing as a 'two-pitch voice'. The only circumstances in which we can expect to discriminate two simultaneous pitches from one generator is if the two component frequencies are widely different in vibration rates with none in between them, as the three-tones experiment showed, and that does not happen with voices; or, that there is a very large component which is not related to the others, and that would not be a basis for relating pitches. If one looks at the sound spectra of sung notes there are large harmonics corresponding to vowels but descriptions of sound in the air do not demon-strate sensations. It is one of the most important things which physiology exposes about physical acoustics. A set of harmonics produces one pitch sensation. That is the most remarkable property of the hearing system, and I still put it in those terms despite having offered a possible suggestion in Chapter 9 as to how it happens.

There is, however, a third argument which in my view is the most powerful. If you have little familiarity with the history of discovering the properties of the physical world and the process of application and invention, it may not appeal to you at once, but if so I hope you would discuss it with any friend who is conversant. If pitched music was invented by the human voice, then we made musical instruments to copy it. If it was invented on artefacts, the voice imitated it. We have no difficulty with the second proposition, because that is what voices do. But the history of invention is accidental discovery, followed by its application if the discoverer had sufficient knowledge and lateral thinking of an appropriate kind to see that it had an application. The ability to invent anything to perform a purpose, requires an accumulated knowledge of the behaviour of the physical world, a knowledge of principles, and that is an advanced form of human achievement. Do you think the first person who broke a flint even said to himself 'I can make axes, spearheads, skin scrapers and knives if I crack these appropriately'? More to the point, did the person who first blew across a panpipe tube think 'That's like the sound Aunt Mary makes when she sings'? In any case it wasn't. It had an entirely different starting transient which is the main identifying feature and it had different tone. It would have had steady pitch, but the recognition of steady pitch is a more sophisticated idea. And even if some early person did think a panpipe was like a voice singing, it would stretch credulity that someone in some other part of the globe thought that a xylophone did. Any theory of the origin of pitched music has to account for the appearance of the pentatonic scale on a variety of different-sounding instruments some of which had very irregular harmonics, in a diversity of places. We don't know and we will never know how such music began. Even if we asked every reader to vote, it would only tell us which is the most convincing story. The importance lies in whether the voice or the artefact then produced the early developments which laid the foundations of tonal music. Communal chanting, humming and primitive dancing, all of which make a lot of noise, developed at some time too, but when? There were less than a hundred humans of all ages in the entire Tehuacan Valley by 7000 BP.

12.3 Memory

How our memory actually functions is one of the ultimate bastions which science has yet to penetrate, if indeed it ever will. We can only describe what it does, and we have to be as careful in describing the circumstances in which we discuss remembering music as we have to be with the use of music terms. I described the earliest stages in the selection of pitch sequences as putting a sensation of a pitch into short-term memory and comparing the sensation of the following pitch with it, because that is what actually happened. The experimenters selected the sensations of successions of fifths, fourths and thirds. I have said it was because they liked them. In the light of the arguments in this book, if anyone can propose an alternative reason for their choice, I would be very interested. Is the reason because in whatever form the first sensation is stored in memory, there is a simple relationship between it and the sensation produced by the second sound? It is the only answer I can advance, and it may very well be true, but that is not far from a long slippery slope at the end of which we may try to say why people like pieces of advanced music, when we know very well that some do and some don't like its sensations. Whether it is about food, drink, pictures or music, the only thing an individual can say is that he or she liked the sensations and not whether anyone else will. But even today, very few people positively dislike a simple pentatonic tune, though they may not find it interesting.

A better approach might be to ask how close to simple ratios were the pitch-patterns of the first selected fifths? In Chapter 6 we discussed the phenomenon of fifthness, which exists from a quarter tone flat to a quarter tone sharp. When we carry out those experiments now, we know all the terms and we know what the sensation of a fifth with pitch-patterns close to a $2:3$ ratio is like. But the sensations would be the same if we didn't. Fifthness would still be there. This seems to me to be a much stronger argument in favour of a cortical activity which identifies a unique relationship of two steady-pitched sound sensations from pitch-patterns even roughly approximating to a $2:3$ ratio, and we suggested how that could come about in the process of matching patterns.

One could take a different line and say that in the iterative process of producing a pentatonic flute, unless one got the fifth ratio right, the error would reflect on all the related ratios after the fashion of the comma in the Pythagorean scale, but I think that is far too sophisticated an approach. It is how one develops a better flute from a basic one. And in all this we must not lose sight of the ability of a wind player to vary the pitch by blowing. That was why baroque flautists nodded like hens. Being an adequate player has always involved an ability to sense simple intervals without any tuition, and as every violin teacher knows, some can and those who really can't never will.

But I do not think it is possible to seek any common basis for musical memory in general, though several people have tried, because there is no reason why any two cortices should store the sensations in a similar fashion or retrieve the information in a similar way. There is abundant evidence that we don't. We don't have any control over how we organise our cortex. We can describe what some people achieve, but not how they do it. If I say that someone remembers a tune as a sequence of pitch-pattern ratios, because when they recover it and play it on a violin or sing it (both processes being able to produce any pitch-frequency in the range), and the pitches are in the same series of pitch-frequency ratios, though from a different instrument to that on which the tune was heard, who is going to argue that that individual didn't store and recover it that way? We can teach young birds to sing tonal tunes by imitating the way they acquire their natural song from their parents (though it would be as irresponsible to release them as it is other things that man has modified).

Remembering something new with immediate replication may be quite different to recognising music from a long-established memory and identifying it. Some people can abstract. Although most of the interest in musical memory seems to have been concerned with pitches, I said in Chapter 11 that the distribution of starting transients in time and the time framework, are just as much an integral feature of a tune as are the pitches. Tapping the distribution of starting transients as a sequence on a block of wood and asking listeners to identify the tune is a well-known quiz game. People find it more difficult if the transients are produced by a computer programme, with identical sounds distributed square, than if tapped by an experienced musician, but that's all part of the way sensations are stored. As I mentioned in Chapter 7, like many traditional jazz players, I can identify a lot of music simply by hearing its harmonic sequence; it extends to music of very different kinds, and I know others who can, depending on their store of music. But I very often don't recall the tune quite correctly. If one is a natural improviser, a tune is something that fits the harmony. Perhaps it is also the reason why real folk-tunes get slowly transformed when transferred by the aural route over centuries. The starting point of musical memory isn't the sound in the air, but the form in which the information *enters* the cortex, and I don't think there is any future in trying to explain musical memory until we know how a cortex works – if ever we do.

There is, however, one facet of musical memory I touched upon in Chapter 1 which has implications for how people learn to play instruments. 'Listen and Sing' is the standard way in which classes of children learn songs at school; the piano plays it and then they sing. It is the standard way the chorus of stage shows learn, and surprisingly, some of the highest-paid opera singers can't read music and have to have the tunes hammered into them by a répétiteur. Most instrumentalists can listen and sing. Remarkably few who have been orthodoxly trained from the beginning, can listen to eight bars of a straightforward tune they don't know (or even one they do and have never played before) and then play it on their instrument. That is the starting point for the individual who discovers how to play an instrument with no tuition. It is called playing 'by ear' and it is a sensible description. The process starts by putting the tune from the ear into the cortex store; then, formulate a pitch in the auditory cortex, instruct the body to play it and monitor that sound from the ear to the cortex again. It is a short closed loop. That is what happened in music for thousands of years. Playing 'by eye', whether sight-reading or something the player knows well, is: notation to eye to visual cortex, instruct the body to play it, and monitor by ear to the auditory cortex. Whether the pitch is correct also requires the visual cortex to have informed the auditory cortex what to expect. It is a very complex process, and it may help to explain why 'eye' players sometimes play mechanically or don't play in tune. One characteristic of the untutored is that they usually have impeccable intonation (and often natural lift too, but that may be part of the motivation for wanting to play). It suggests that the two procedures involve different and in some people incompatible cortical memory and processes. And it might suggest that there is a good reason for children to begin without notation, as for example in the Suzuki method.

12.4 Other pitch systems

I have limited the discussion in this book to western tonal music and especially to the simpler forms of it. It is the form of music with which I am very familiar. In its simpler forms it does appear to have immediate appeal to very large numbers of people in most countries into which it has been introduced, regardless of their traditional musical cultures. It is tempting to suggest that that is because they all have the same hearing machinery, and that machinery can present its simple logic. Does it necessarily follow that sensations which are based on an inherent logic are attractive, even though the logic is not recognised at the conscious level? Music of most kinds and cultures is readily available through radio today. We know from the experiments in the twentieth century that non-tonal music based on the western twelve pitch scale has very limited appeal, even to those who enjoy very complex tonal music. But the traditional music of some cultures using different pitch systems and varying pitch does appeal to some whose main experience has been of western tonal music. Whether varying-pitch music originated with the voice or with variable-pitch instruments, whether it has a simple logic such as appears to underlie the tonal system, it is for others to demonstrate if they wish. What I have read about, for example, Indian music, is as incomprehensible as is the orthodox terminology of western music as a basis for discovering anything about it in the way that we have investigated tonal music in this book. For one cannot describe a sound sensation. One can only describe what may produce one. And a lot of our orthodox terminology does not even achieve that.

12.5 Believing is hearing

If everyone obtains the same sensations from the same sounds, it is difficult to understand how some people can hear things that they can't. Mersenne (1630) said he could hear all the harmonics of a note and believed anyone could do so (if the translator interpolated his writing correctly). The scientific evidence is entirely against this, but how does one test what a person imagines? I am sure people who know about harmonics can imagine a pitch which fits with a harmonic of an instrumental note; in fact one does not need to know about harmonics to do so, for the very essence of the argument throughout this discussion is that in creating simple music, composers could imagine pitches with rates which made simple intervals, and which will happen to correspond to harmonics. The critical test would be for claimants to *demonstrate* that they could identify one or more harmonics of a sound which consisted of several unrelated ones. On the other hand, there are plenty of instances of the impossible; there is the well-known case of the sound of old violins which has been adequately rebuffed in blind tests, and I mentioned the cellist who heard steel strings as gut. There are those who believe they hear the pitch of G♯ to be different from that of A♭ when it is the same note played on an equal-tempered keyboard instrument. Examples of people hearing differences in intervals when the actual pairs of pitches lie within the pitch-bands of the sounds and cannot be discriminated by hearing are legion. The most recent one of which I have read is a man who heard beats from thirds played on a normal piano. In almost every case the belief is based either on something they have been told by someone else who so believes, or something they have read and misunderstood. There doesn't appear to be an exact parallel in any of our other senses. If people continue to believe that they can hear things after one has explained to them why they can't, all one can say is so did Joan of Arc, and look what happened to her.

12.6 Pleasure and wonder

Towards the end of Chapter 1, I suggested that music was unlike any other human discovery and invention. Unlike other sensations we get, for which we can propose a good biological purpose, the only reason we can suggest for music is that we selected the sounds because we liked them and we liked the sounds because they were the ones we selected. It is a circular but incontrovertible argument, though probably from the very beginning, it was a small number of people who selected them, and a larger number of recipients who decided whether they liked them, and of no age is that more obvious than today. It became associated with many kinds of social activity, but I can think of no other basis for the discovery and development of instruments, or for composition, than that people found the sensations pleasurable, though later they may have begun to associate them with pleasurable activities or with emotions. There was and is no need for anyone to have the slightest idea about what the sounds actually are, how they are produced or how we hear them, to compose, play or listen. It has always seemed some kind of magic.

But anything which starts as magic and wonder is bound to invite the interest of inquisitive man, a better name for him than *sapiens* meaning wise, and attempts to discover the nature of music may pre-date the Greeks. In classical Greece, studies of natural events were a part of philosophy. In later history, the course of inquisitiveness diverged. On the one hand it gave rise to an enormous body of literature concerned with the wonder, some of which might generously be called the philosophy of music. On the other hand, it was subsumed in the branch of knowledge which was called natural philosophy for a very long time, and we now call the physical sciences, from which arose the science of acoustics. But because of the complexity of the living system, in which the senses are especially difficult to investigate, the body of knowledge essential to begin our quest to understand music had to wait until the twentieth century. Yes, I say begin, because I regard the empirical discovery of the whole of music, and the science of acoustics, as the preliminaries. Music is a phenomenon of hearing. I have given a very simple version of an exceptionally complex machine, and restricted it to the features which appear to relate to music. Much more detail is known, and there is much more to be discovered about the machine. The auditory cortex lies beyond that.

Appendices

Appendix 1. Interval names

Intervals are often defined as the distance between two notes, which is meaningless, or the difference in pitch between two notes which is wrong. Neither suggestion conveys any concept of what an interval is, or how we recognise one. The archaic terminology is based on a visual image of a seven-pitch major scale, incorporating a mystical element by calling the octave, fifth CG and fourth CF 'perfect' intervals, but the second CD, third CE, sixth CA and seventh CB 'major'. This was adapted piecemeal when the five additional pitches created the twelve-pitch scale. There is some excuse for calling CE♭ a minor third, but it is meaningless to call every interval minor when reduced by a semitone, and irrational to call all intervals diminished when reduced by two semitones, or augmented if they are raised a semitone. I don't know what answer teachers give if someone dares to ask why? As Table 2 shows, the white-note scale contains every interval of the twelve-note scale.

Appendix 2. Selecting pentatonic pitches

Starting with a C pipe, Table 5 shows that any of the pitches E♭, E, F, G, A♭ or A make a simple ratio with C. The first added pitch restricts what further pitches satisfy the requirement of making acceptable successive intervals with all those already chosen. Seconds created in this way are acceptable in succession, if they can be related to the other pitches already selected. The process always ends with CDEGA, CDFGA, CDFGB♭, CE♭FA♭B♭ or CE♭FGB♭, all versions of the same pentatonic scale. No further pitch can be added which relates to all the others.

5th	4th	ma3rd	mi3rd	6th	aug5th	mi7th
2:3	3:4	4:5	5:6	3:5	5:8	4:7
CG	CF	CE	CE♭	CA	CA♭	CB♭
DA	FB♭	E♭G	EG	DB♭		
E♭B♭	DG	E♭A♭	DF			
	EA	GB♭				
	FA♭					

Table 5. Approximate ratios of pitch-frequencies of all the pentatonic scales starting with C, apart from seconds.

The sequence CGDAE derived by exact ratios is $3/2 \times 3/4 \times 3/2 \times 3/4 = 81/64$. No musician derives E from C that way. The ratio used is $80/64 = 5:4$.

Appendix 3. An alternative dodecaphonic route

The organ, possibly invented in Egypt and entering Europe through the eastern Mediterranean, led painstakingly to the twelve-pitch scale, but there is a rapid way in which that scale was discovered, which entered Europe from north Africa at the western end of the same sea. It also depended on producing fixed pitches with good harmonicity, from the plucked vibrating string. The principle is very simple. Suppose one experiments with an instrument of the lute family such as the 'ud. Tune five strings to the pentatonic scale CDEGA. Extend the pitches by placing a gut ligature fret round the neck raising the E to F, making acceptable intervals with C, D and A. It produces a new pentatonic scale: CDFGA. But the same fret across the neck will produce C♯, D♯, G♯ and A♯ (or their enharmonic equivalents) on the other strings. The fret added to produce F creates every pitch of a twelve-pitch scale except B and F♯. One further fret and the instrument produces all twelve pitches, almost an instant recipe. Unlike the laborious discoveries on a keyboard instrument, the problems of the scale are revealed immediately. If one positions frets to produce the simple intervals for one or two heptatonic scales, they do not create pitches which make simple intervals in the other heptatonic scales which can now be played. The same problem occurs with fretted viols which were developed at a later date.

The inference from the data gathered by Helmholtz (1870) and collated by Ellis, is that by the twelfth century an Arab culture was not only making good enough strings and refined tuning mechanisms on a lute type of instrument, together with inlaid frets, to appreciate the differences between the intervals produced by pitches determined by such a fret system, but that they found an ingenious way of changing the pitch of a stopped string while playing; it parallels the split G♯/A♭ key on an organ. This was two very short inlaid frets side by side under a string, one slightly further from the nut than the other, so that by moving the string sideways with the finger it could be stopped by one fret or the other and sound a slightly lower or higher pitch as required. Fretted instruments had a huge influence on the development of Hispanic music. The elegance of the split fret was replaced by intermediate fret positions producing equal-temperament on the modern guitar, but the traditional 'ud or oud has frets for a variety of intervals.

Appendix 4. Pitch discrimination and intervals

Our ability to distinguish that two pure-tone pitches are different can be assessed in one of two ways. Subjects hear two tones in succession and say whether they are the same or different, or they listen to a tone which is continuously changing in frequency up and down twice per second and say when they can hear it is changing. The two tests give very similar answers. The

Frequency (Hz)	31	62	125	250	500	1k	2k	4k	8k
A	2	3	4	4	4	5	7	10	25
B	1.8	3.7	7.5	15	30	60	120	240	480

Table 6. Frequency discrimination. A: Difference in frequency between two pure tones which can just be detected. B: Difference in frequency between two pure tones a tempered semitone apart. To use the Table: at around 250Hz (about middle C) we might just distinguish 248Hz from 252Hz, differing by 4Hz. A semitone at 250Hz is a difference of 15Hz between two pure tones. So we can just distinguish 4Hz in 15Hz, or very little better than a quarter of a semitone in pure tones.

	A	B	C
2nd	8 : 9	− 0.04	8 : 8.98
minor 3rd	5 : 6	− 0.16	5 : 5.95
major 3rd	4 : 5	+ 0.14	4 : 5.04
4th	3 : 4	− 0.02	3 : 4.004
fifth	2 : 3	− 0.02	2 : 2.997
aug. 5th	5 : 8	− 0.14	5 : 7.94
6th	3 : 5	+ 0.16	3 : 5.04
minor 7th	5 : 9	− 0.18	5 : 8.91

Table 7. A: The simple ratios of just-temperament intervals. B: How much equal-tempered intervals are greater (+) or less (−) than just ones in fractions of a semitone. C: The ratio of the equal-tempered interval. B represents the disparity hearing would have to judge to distinguish consecutive intervals of the two kinds and C the disparity to distinguish simultaneous intervals. It is doubtful if listeners would distinguish either when typical instrumental music is played. Under test conditions listeners may be able to distinguish some from a small harpsichord. Since the semitone, the augmented fourth and the major seventh do not have a fixed simple ratio, one cannot give values for discrepancies.

values in Table 6 are probably the best that people can achieve under laboratory conditions.

With monophonic instrument or piano sound, if the harmonicity is good, the discrimination of pitch-frequency may be improved by the contribution of the higher harmonics to pitch, but it will be broadened by timbre. Under practical conditions it is doubtful if we can ever do better than a tenth of a semitone, until we hear pitch-frequencies above treble staff (1kHz upwards). If the harmonicity is poor we shall not do as well as that. In a consecutive interval we have to remember the first pitch-frequency and compare the memory with the second pitch-frequency. We judge a simultaneous interval in an entirely different way as shown by Table 7 (see also Section 6.12 and Chapter 9).

Appendix 5. Tuning a keyboard by beats

There are a number of formulae for producing equal-temperament by beats; a standard method is to set a groundwork octave by obtaining the correct rate of beating between fifths and fourths and to check them with thirds. Two complications are involved. Suppose we want to narrow a perfect fifth with pitch-frequencies of 100 and 150Hz by making 150 into 149. When there are no beats, the third harmonic of 100Hz is 300Hz, coinciding with the second harmonic of 150Hz which is 300Hz. We lower 300Hz to 298Hz, by obtaining 2 beats a second, which makes the 150Hz into 149Hz. To change the 100Hz, the beats must be three times the rate at which the 100Hz is to be changed. Similarly, for a major third, ratio 4:5, the beat rate must be four times the amount by which the higher string is to be changed, or five times the amount by which the lower string is to be changed.

But the rate of vibration of all the strings increases as one ascends the octave. The rate of beating which changes an interval by a given fraction is nearly twice as fast at the top of the octave compared with the bottom of it. The groundwork scale is set on single strings because only with them can one hear beats at all, then the unisons are tuned and everything else is done by octaves. A tuner thumps the keys when tuning to hear the beats but it also helps the string to slide over the bridge so that the tension is the same on both sides of it. This is essential if a piano is badly out, like one I tuned in a pub miles from anywhere, near where I was stationed in 1942. Even then the tuning needed trimming after I'd played it the following day.

Appendix 6. The abominable cent

There are several reasons for abandoning the unit called the cent. It is worth reading this paragraph even if you have never heard of it, or are like those music students who failed to understand it when it was explained. If people express intervals in cents, it suggests that they are thinking about them in the wrong way. The cent was actually invented 'so that intervals could be added together' (Wood, 1961), thereby perpetuating the attitude that a fifth plus a fourth makes an octave, which was criticised on page 1.

A cent is not just one hundredth of an equal-tempered semitone, but an equal-tempered one hundredth. There are twelve such semitones in that scale, so there are 1200 cents in an octave; an equal-tempered major third includes four semitones and is 400 cents. We cannot hear a pure tone to better than 4 cents at the most discriminating part of our hearing, around 2kHz; for most of the pitch-frequencies used in music we can do no better than 10 cents with pure tones. Even if one assumed that monophonic instruments produced the same frequencies every time, good scientific practice would require an interval to be stated with limits, such as 400 ± 5 (between 395 and 405 cents averaging 400). Quoting a pitch in cents gives a completely false impression of how accurately we can hear or play, and, frankly, how accurately someone thought they could measure a pitch-frequency. Sounds should be described as frequencies, which is all that can be measured. If the purported pitches of an ethnic instrument are in cents, and one wants to discover whether they approximate to any simple ratios, one has to convert them by a formula and a book of logarithmic tables. Even then, cents do not tell one frequencies, or whether the instrument was a penny whistle or an alphorn and some authors do not include that clue.

While I regard Helmholtz as the greatest figure in the investigation of music by the application of science, the mass of information on primitive scales in the appendices of his great book are given in cents. He could not have measured them that accurately. And he added, when discussing some of their users that 'they had very bad intonation' which, to say the least, is a typically Victorian attitude towards the uncivilised.

Appendix 7. Repetition rates

The chord example [CEG] which we have represented as 400:500:600 has a chordal repetition rate of 100 times a second, but the recognition of a major triad is the relationship of the chordal repetition rate to that of the components: one quarter the rate of the 'C'. If we spread the notes and put C in the bass, to obtain 100:500:600, the repetition rate is still 100, but if we put G in the bass, as 150:400:500 the rate is 50, and with E in the bass as 125:400:600 the rate is 25. As musicians know, using inversions in bass parts does produce recognisably different sensations, but sensations which continue to be identifiable as 'major', because of the simple relationship of the repetition pattern to the repetition of the components creating it. Each kind of chordal sensation which can be named: major, minor, dominant seventh and so on, has a different repetition pattern related to the repetition rate of its components. There is little point in demonstrating this in detail, because there is a simple argument applicable to all sound sensations; if the patterns were not similar musicians could not recognise the similarities, and if they were not different we could not distinguish and identify the differences. The patterns are the only things which provide the similarities and the differences. This supports the premise in Chapter 1 that the sensations of steady-pitched sound has properties different to all other noises, for noises can only be individually identified. Presumably it is because the process of identifying intervals and chords is by recognising likenesses rather than differences, that most people don't do it.

Appendix 8. Anatomical evolution

For practical reasons, the sense organs and limbs of most animals are symmetrical about the mid-line, and the parts of the brain which receive signals and give instructions automatically are correspondingly symmetrical. The cortex also consists of two halves, but it is far from symmetrical in its functions. Different things are dealt with in either half, and they are joined by massive nerve cables which co-ordinate the two sides. One possible reason why

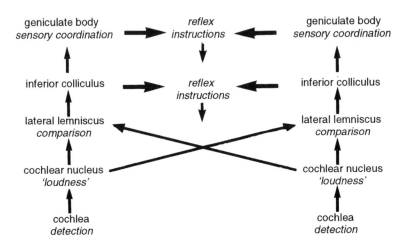

Fig. 25A. Hypothetical primitive sound-detecting system. This could distinguish real events and make a simple assessment of loudness. By comparing signals from both ears it might produce a simple sense of direction and, combining this with the input from other senses, produce automatic reactions to sound. Compare Fig. 25B, and see Appendix 9.

individuals may have very different abilities in respect of language, pitch and rhythm is that they may each be functions in different parts of the cortex, and for playing music the mechanical activities have to be co-ordinated with all of them.

Figure 25A is a hypothetical primitive auditory system consisting of a pair of vibration detectors (ears) and processors, pathways to both sides of the system meeting in co-ordinators, and thence to an executive unit. A processor received pulses from a detector and determined whether they arose from a real event (see Appendix 9) and how big the event was, the equivalent of loudness. The co-ordinator compared events on the two sides, a normal feature of paired sense organs. The executive unit initiated immediate automatic responses to sounds, by sending signals into the nerves in the spinal column controlling the body, summarised in the three possible reactions: freeze, flight or fight.

Figure 25B is a block diagram of the advanced auditory pathway in higher mammals including ourselves. It may appear complicated, but it looks far more so when one works it out with a real brain! And if you like designing model railway systems, you can develop it from the system in Figure 25A. Duplicate the connections between the primitive processors and the co-ordinators as parallel routes. Expand them into new stations (the olives) half way along those routes. That is how brains evolve. A direct route, which might signal 'Vibration!' is still there. It might still do so. The layout of nerves in the inner pair of olives suggest that they are comparing the signals from the two sides in detail. The outer pair may be comparing loudness. The co-ordinator combines all this information into clarified signals which determine sound

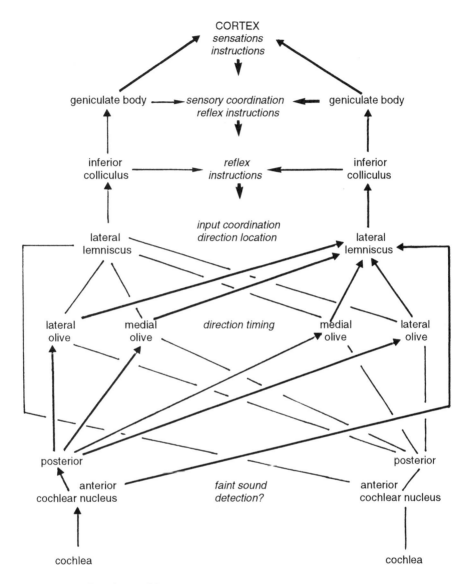

Fig. 25B. The advanced hearing system. Compare Fig. 25A. Four processors (the olives) have been developed between the cochlear nuclei and the lemnisci. The medial olives are believed to assess time difference and the lateral olives probably loudness difference between the sound at the two ears. The cortex receives the processed signals as sensations. See Appendix 9.

direction, and that can be used by the executive unit to give automatic instructions. The cortex then receives the signals which have been processed for those purposes, and they produce the sensations. Deciphering the pulses in individual nerves at the level of the olives has proved almost impossible, and in the cortex totally so. There are also some nerves running in the opposite direction throughout the system. They may, for example, enable the cortex to over-ride automatic reflex responses to sounds.

Appendix 9. The primitive processor?

The cochlear nucleus is the first unit of the auditory pathway to which the signals from the ear are passed, and it has two distinct parts. Each of the seven thousand or so nerves from the cochlea divides when it gets to this unit. One branch of each enters the posterior part, and they split into the four routes to the olives, transmitting the details of the pulse trains. The other branch of each goes to the anterior unit. Its structure suggests that it may have been the primitive processor, carrying out the two basic functions, to determine that pulses from the sound detectors meant a real event, and if so, to give a crude indication of how intense it was. A problem in all nervous systems is that the more sensitive nerves and sensors are, the more likely they are to fire accidentally rather than be fired by a real event; mouse traps are similar. There is spurious firing, usually called noise (yet again), in all nervous systems. If the vibration detectors are sensitive enough to detect tiny vibrations, some of them will fire spontaneously. How does the system know which firings are a real event? The solution is called coincidence counting, and employed in such fields as radio astronomy, but animals discovered it millions of years ago. A simple analogy explains it.

A spy is to signal from a distant mountain, by lighting one match at the darkest time of night; it can just be seen as a flash by observers far away. If anyone looks intently into complete darkness, they see random flashes because the sensors in the eye can fire spontaneously. One observer cannot tell a real from a spontaneous flash. But if two observers each call out whenever they see a flash, then if they call out together they may have seen a real flash. The more observers who call out simultaneously, the more probable the event is real.

The anatomy of the processor was described by Lorente de No (1933) and an interpretation of how the nerves are connected in it was suggested by Whitfield (1967). Each nerve from the cochlea divides into two hundred or so fibres. These make junctions with receiver nerve cells; there are as many receivers as there are incoming nerves. They are connected such that each receiver can obtain pulses from a different sample of two hundred of the incoming fibres. A random sample will include fibres from different regions of the cochlea. If a receiver is set so that it fires if it receives a certain small number of pulses all within one millisecond, this is a coincidence counter, and the larger the vibration, the more receivers will fire, giving a crude indication of the intensity of the vibration. Richard Pinch kindly ran a computer programme in which we arbitrarily put the number of simultaneous pulses required to fire a receiver at five. It showed that when any 40 of 7000 sensors fired, one receiver would fire, and when about 400 sensors fired, all 7000 receivers fired.

Obviously such a system would completely confuse any information about sound frequency which might be present, but detection and a crude indication of intensity was the first priority, and when there was a very poor cortex it could make nothing of frequency information even if it received it. Whether this most primitive part of the auditory pathway now acts as a detector of minimal vibration, or like many vestigial bits of our bodies it is no longer used, it does not interfere with the transfer of pulses carrying very detailed information about received vibrations along the other routes to the advanced processors and cortex.

Appendix 10. The ear structures

Since the air-filled cavity containing the three tiny bones called malleus, incus and stapes is sealed by the eardrum, the pressure of the air must be the same on both sides, or the drum will be tensed and not vibrate freely. The Eustachian tube runs from the cavity and opens into the throat, and when we go up a mountain or in aircraft into lower air pressures, air leaks down the tube, keeping the pressure on both sides of the drum the same. On returning to high pressure, air does not go back as readily, and we are often slightly deaf; swallowing may open the lower end of the tube. The main role of the bones is to change the form of the vibration in the following way. Sound in air is a large vibration, comparatively speaking, at low pressure because a gas is easily compressed. Sound in a liquid such as in the cochlea, is a small vibration at high pressure, because it needs a much larger force to compress liquids. The amount of energy in the sound is little changed.

Most of the sound energy is used in vibrating the basilar membrane. There is a very thin membrane called the oval window between the cochlea and the middle-ear cavity which lets any unused sound pass out and be absorbed on the soft walls of the cavity. It is hardly picked up at all by the three tiny bones, for otherwise, unlike the popular tune of the 1930s, the music would go 'round and round' and produce continuous echo. In the recent gimmick called 'surround sound', discussed at the end of Chapter 10, that is just what the electronic trickery makes the sound do.

The story of the three tiny bones is more entrancing than any fairytale about the origin of music. When our remote ancestors crawled out of the sea, their legs would not support their heads off the ground; crocodiles still have their legs at the wrong angles to lift the body up. These ancestors' lower jaw consisted of four bones on each side, and the jaw rested on the ground. The bones picked up any vibration of the ground, and three of them conducted the vibration to a very primitive detector in the skull. Our jaw has only one of those bones on either side, and the same three other bones became the ones in our middle ear – still conducting vibration to the detector.

Appendix 11. Hair cells

Why there should be two sets of cells has been a puzzle from the beginning of speculation on how an ear worked. Both sets have minute 'hairs' on their ends, and no one suspected that they worked in entirely different ways, but biological systems often adapt cells already present for new purposes. The discovery that one set expands and contracts is so recent (see, for example, Duifuis *et al.*, 1993; Yates, 1995) that its role in any general theory of hearing is in its infancy. It may have important implications for speech reception. I am not aware that anyone has suggested in as many words that the system does apparently separate pitch from loudness in hearing music, though how we hear music is a very minor part of studying hearing.

The model illustrated in Fig. 10C does not preclude the possibility that at high sound levels, the vibration of the membrane could tilt the triangular frame directly rather than via the expansion of the triplets. It does not affect the argument; what matter is that the cells and frame are vibrated at sound frequency. We have fewer 'amplifier' cells at the two ends of our basilar membrane (Wever, 1949). That might correspond with our poorer hearing of low-level sound at the two ends of our frequency range. Our anatomy remains the same however much our ideas on how it works may change.

Appendix 12. Recording auditory nerve pulses

In their classical experiments, Kiang and his colleagues (1969) investigated the cat, while Rose and his co-workers (1967) used spider monkeys. They were able to insert minute electrodes into single nerve fibres of auditory nerves and record the patterns of impulses from a single hair cell while various pure tones between 50Hz and 2.5kHz were played into an ear (2.5kHz was as high a frequency as it was technically possible to use). In these experiments, over a thousand successive pulses were recorded at each frequency. What they actually recorded was the time interval between each pulse and the next, in relation to the period of the pure tone. One of several difficulties about such technically demanding experiments is that one can only determine approximately where the hair cell is in relation to the nerve fibre from which one records. The experiments show that a particular hair cell codes sound over quite a range of frequencies; the level of pure tones used was fairly high (see also Appendix 17).

Appendix 13. The irregular distribution of pulses

Table 8 gives approximate values extracted from the recordings of Rose *et al.* (1967) of the distribution of pulses in an auditory nerve fibre for various pure-tone frequencies stimulating the ear. In line A, 1–7 indicates that the gap between successive pulses may be any number of periods between 1 and 7. As line B implies, the largest number of pulses are spaced at one period, the next largest number at two periods and so on. The longest gaps are rarest. The

Frequency (Hz)	100	120	250	400	1000	1700	2000
A. Number of different periods	1–3	1–4	1–7	1–10	1–17	2–18	2–18
B. Average gap between pulses in periods	1.25	1.4	2.4	2.9	4.3	4.1	4.2
C. Pulses per second	80	85	104	137	232	414	476
D. Average time between pulses (msec)	12.5	11.6	9.6	7.2	4.3	2.4	2.7
E. Period (msec)	10	8.3	4	2.5	1.0	0.6	0.5

Table 8. Distribution of nerve pulses. See Appendix 13.

shortest gap at 1700Hz is of course two periods. Clearly, the average time between pulses in line D cannot provide the period of the vibration given in line E. If we divide the frequency by line B we have the number of pulses per second produced by a signalling hair cell in line C.

There is ample evidence in Chapters 9 and 10 that sound sensation is created by the periodicity of pulses. But there is no direct evidence of the form of the nerve pulses for frequencies above about 2.5kHz. The very highest pitch-frequency ever used in music is 2.5kHz, E above the C three octaves above middle C. Just as we experience no discontinuity in a gliding tone around 1kHz, corresponding with the maximum firing rate of nerves, so also there is no discontinuity around 2.5kHz, and our discrimination is excellent at 4kHz. It would be remarkable if the hearing system used some completely different way of signalling the pitch of high frequencies.

Appendix 14. The length of resonant vibrations

The actual vibration of the resonating membrane is minute and excessively difficult to observe directly. And since the vibration decreases to zero at either end of a pure tone resonance (Fig. 12A), it is impossible to measure its length. The main source of information is provided by looking at the mechanism the other way round, that is to say, the range of frequencies over which an individual hair cell signals. At reasonably high sound levels, if, for example, one hair cell can signal pure tones from 50 to 500Hz, then we know that the resonant lengths of loud pure tones from 50Hz to 500Hz extend over that hair cell. If another cell only signals from 2 to 4kHz, then those vibrating lengths extend over that cell. If the word 'only' seems odd, the spacing is somewhat like a piano keyboard; 50–500Hz is more than three octaves whereas 2–4kHz is only one octave. A large number of measurements of this kind have been made, for example by Palmer and Evans (Palmer, 1995) on the cat, and Professor Palmer has kindly given me permission to use them, from which I have extrapolated the lengths of vibrating membrane used in Figs 21–4. It confirms that low frequencies vibrate a considerable length of membrane and their vibrations overlap considerably, high frequencies vibrate much smaller lengths of membrane.

Palmer's results also show that as the intensity of the sound increases, the

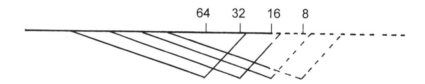

Fig. 26. Basilar membrane vibration at very low frequencies. As a pure tone frequency drops from 64 to 8Hz, the resonant envelope moves to the end of the basilar membrane, and then in theory beyond it as shown by the dashed lines. The end of the membrane continues to vibrate the hair cells there, but at some frequency below 16Hz there is no longer a continuous input from the cells as the travelling wave sweeps along it. The listener then receives an intermittent sensation at sound frequency as explained in Chapter 10.

frequency band over which one hair cell will signal widens. At low sound intensity, only cells in the region of maximum resonance signal, and as the sound increases, more and more cells at the extremities of the resonance fire. A somewhat ambiguous term is used for this; it is said that the hair cells are 'tuned' to the frequencies to which they respond. Not only will musicians smile at the idea that something is tuned – to all the frequencies between 2 and 4kHz at once, but it would appear to be the membrane which is tuned rather than the hair cell.

A further piece of evidence is provided by very low frequencies. As the frequency falls, the resonant length moves towards the end of the basilar membrane, and by 16Hz, it will have reached the end of the membrane. If the frequency is lowered to 8Hz, the end still resonates though it is not long enough to include the whole of the Bekesy wave. There are parcels of continuous pitch sensation interspersed with silences at 8 per second. The explanation is, presumably, that a single 'down' part of the resonant wave appears, runs off the membrane and then reappears (Fig. 26). This provides further support for the inference that we need a continuous supply of pulse trains from several hair cells to obtain a pitched sensation of any kind.

Many of the anomalies which have been discovered in the hearing system occur when extremes of high and low level sound, and very low and high frequencies are used. One may learn useful things by testing physiological mechanisms in that way with unnatural sounds. What is relevant to music must be limited to the circumstances in which we hear it.

Appendix 15. The place theory

The idea that sound frequency was signalled by *where* the basilar membrane vibrates, rather than *how* it vibrates, was called a 'place' theory: the place where sound is detected. It presumed that the brain created a map of the cochlea by learning, in order to know what nerve indicated what sound frequency. It was the pre-1939 view of the senses that the cortex sorted everything. It was not unreasonable then. Also, the brain does have a general map of our body in that we know how our limbs move, because each limb muscle has sensors which signal where it is. When electrical pulses in large nerves were first detected, it was realised that nerves could signal many things including possibly sound frequency, by a code of pulses. There followed a continuous argument between the proponents of the two theories, which is set out in the *Theory of Hearing* by E. G. Wever (1949), an example of the finest accessible scientific writing, at the time of great scientific developments in physiology, in which I was fortunate to take part.

No one has attempted to show that a place theory accounts for automatic processing, or for the detail in which the sound sensation is transmitted to the cortex. The pulse code principle outlined in Chapters 9 and 10 does provide a cohesive explanation for musical phenomena. We do not, so far as I know, have any direct evidence of the form of the pulse codes above about 2.5kHz. It is unfortunate that people who believe we have to learn pitch and create a map in the cortex, and that the twelve-pitch western scale is imprinted by brain-washing, have jumped on the lack of direct evidence of coding above 2.5kHz to advocate that this must happen at all sound frequencies. As we see in Chapter 10, we *do* have a map of a part of the two cochleas in the brain, but it is inherited not learned, it isn't in the cortex, it isn't used to create the pitch sensation – and it only covers our lower frequency range up to about 2kHz.

Appendix 16. Loudness

Table 9 gives approximate values for the comparative sizes of the vibrations of pure tones which are judged to sound equally loud, for three different levels of loudness, which might be roughly matched with the three levels in music dynamic. For example, at *mf* a pure tone of 62Hz must be twelve times as big a vibration as one of 1kHz to produce an equally loud sensation. It is doubtful if any single unamplified instrument except brass and organs can produce a harmonic louder than *mf* in this table. The purpose of the table is to show the huge difference in the size of equally loud vibrations at low levels of sound, and the much smaller difference at high levels of sound. Since all the harmonics in a sound contribute to its loudness, the higher harmonics contribute most of the loudness of normal notes.

The only physiological basis we have for the general sensation of increasing loudness is that the cortex receives increasing numbers of pulses. Loudness is

	31Hz	62Hz	125Hz	250Hz	500Hz	1kHz	2kHz	4kHz	8kHz
fff	500	250	200	200	100	100	100	100	400
mf	450	120	80	40	20	10	10	10	80
pp	300	100	30	10	3	2	1	2	10

Table 9. Comparative Loudness. See Appendix 16.

determined by the number of cells vibrated, the number which fire, the rate at which they are scanned by the vibration and the average periodicity with which they fire. Let us suppose, purely for discussion, that a pure tone of 100Hz can vibrate 800 signalling hair cells at maximum loudness, and that one of 1700Hz can vibrate 300 cells. At 1700Hz the maximum number of firings, if each cell fired each time it was distorted, would be $1700 \times 300 = 510{,}000$ firings per second, but from the Table in Appendix 16, the average firing would be at 4.1 periods, so that the maximum rate would be about 128,000 firings per second. If loudness is produced by the rate of firing, there is no difficulty in imagining that a scale of 300 steps of loudness could be devised with such a maximum number of firings per second. On a similar basis, there would be a maximum of 64,000 firings per second at the lower frequency to provide for 100 steps of loudness where in musical terms our loudness discrimination is poorer.

The dynamic scale used in music is a different matter. We are not judging 300 steps of loudness in pure tones, each marginally greater than the previous one. We are probably limited to fewer than a dozen steps, judging that one complex sound is about twice as loud as the previous one.

If it is possible for one pure tone to send more than 100,000 impulses a second into the auditory pathway at maximum loudness, then if one listens to very loud music of many frequencies, the number going in, and presumably into the cortex, is frighteningly large – and that is the feeling some people obtain from it. Some years ago, according to one book on acoustics, it was suggested by information scientists that the maximum rate we could take in information through one of our senses was 25 bytes a second (quoted by Hall, 1980). I think the figure needs to be considerably up-rated. There are many recorded examples of the astonishing rate at which people can absorb information, though mostly in relation to visual images and autism. The most spectacular musical feat of which I know, attributed to a presumably normal person, was that of Glazunov who is said to have written out a full score of Tschaikowsky's *Romeo and Juliet* overture almost exactly, after hearing it twice, because he could not obtain the parts to perform it. It makes the ability of musicians like myself, who could play any 1930s dance music on a piano after hearing it once, seem rather trivial.

Appendix 17. Memory storage and recovery of pitches

If loudness is based upon adding pulse trains which are indistinguishable from each other: statistically identical, there is an amusing possibility which I include only as a thought for anyone who studies musical memory. The commonest operation in any nervous system, and certainly in the cortex, is for a nerve to divide many times, each branch making a junction with another different nerve. One pulse stream can thus be transmitted through several nerves simultaneously. Suppose a stream of pulses in one nerve is transferred to two nerves but a time delay introduced into one transfer; it could readily be achieved by having additional junctions in one nerve's path. That produces two identical pulse trains, but the individual pulses no longer coincide in time. For purposes of creating pitch, they are just like two independent pulse streams. A few delayed pulse streams would make it possible to store a pitch as a single pulse stream, but recover it and reconstitute it as a pure tone sensation by such reduplication.

What I find intriguing, though this book is not the place to pursue it further, is that I am sure I auralise themes and harmony in 'pure pitches' in the initial stages of writing a chamber work, rather than doing so at first in instrumental sounds, unless I need to write something like brass fanfares. Leave aside whether you, the reader, do any composing in your cortex; in what form do you auralise a tune? In pure pitches or as you would hear it played by the instrument with which you may associate it? If you can auralise an orchestral work, what type of sound is it in? Considering his frequent disregard for whether singers and instruments were capable of playing what he wrote when he did, I have always believed that J. S. Bach auralised in pure pitches, and allocated the notes to whatever combination of instruments he had available. I don't know whether modern composers auralise, but they also often appear to allocate notes with a disregard for the performers.

Appendix 18. Signalling direction

Brian Moore of Cambridge University, who has contributed so much to our knowledge of hearing through black-box experiments, made a remarkable discovery when listening to white noise (Moore, 1995). White noise is the sort of frying sound one can obtain between broadcasting bands on short-wave AM radio, and is actually a sound of all frequencies occurring randomly and constantly changing. Moore found that if one plays the same white noise into both ears, but delays one by the order of time it would take for sound to travel through the air between the two ears, one can also hear a faint sound with the period of the time delay, that is to say, if the time delay is 0.5msec one hears a faint 2kHz sound. Now white noise may be random but it is still sound made up of frequencies and if there were no frequencies, no vibrations, there would not be any sound or sensation. But it is the ultimate transient noise. We know the automatic processors must be able to determine the time difference, and an obvious way of logging it would be for the pulses representing the sounds to generate the time difference as periodicity. That could communicate the time difference and thereby the direction, both to the automatic executive system and to the cortex.

There is absolutely no reason why, under normal circumstances, we should obtain these signals as a sound sensation. They would be communicating the sensation of direction. We don't hear anything of the signals which tell us we are moving our arm, but there are detailed signals coming back from mechanical sensors in the muscles indicating that every movement is happening.

It is of course possible that the neural difference tone is generated in the same way, but as an accident arising from the functional structure of the medial olives, for there is no biological reason why it should happen. And it also only appears to happen with pure tones up to about 2kHz.

Appendix 19. Words and scents

We may be impressed by the number of word noises we can recognise but it is trivial compared with the recognition of scents by many animals. It is believed to be by sensors, each of which reads a particular shape of a part of a scent molecule, rather in the fashion of trying one side of several pieces of a jigsaw puzzle into one position, and signalling when a piece fits. A surprisingly small number of different kinds of sensor are required, because the small molecules have characteristic shapes. With two sensor types, X and Y, X can signal, Y can signal, or both can do so. They detect three things: x or y or both. Three sensor types detect seven states ($2 \times 2 \times 2$ less 1 when they are all 'off'). Twenty-four types of sensor: 2 multiplied by itself 24 times means that over sixteen million different mixtures of odours could be detected. Animals like dogs have very large numbers of several kinds of sensor, and the number of each kind signalling, indicates the concentration of each chemical shape in a scent. But little is classified. Each scent has to be matched with memorised signals in the olfactory part of the cortex. Similarly, many people can recognise more than half the 100,000 different noise patterns of the words in a word-processor dictionary. But recognising intervals? There are only twelve of them in an octave and a surprisingly small number of commonly used chords.

Bibliography

Apel, W., in *The Harvard Dictionary of Music* (London, 1946) pp. 148, 335, 681

Bachmann, W., *The Origin of Bowing* (Oxford, 1969)

Baines, A., *Musical Instruments through the Ages* (Harmondsworth, 1961) pp. 200ff

Barbour, J. M., *Tuning and Temperament, a Historical Survey* (East Lansing, 1951)

Beament, J., 'The Biology of Music', *J. Psych. Music* 5 (1977), 3–18

Beament, J., *The Violin Explained* (Oxford, 1997)

Bekesy, G., *Experiments in Hearing* (New York, 1960)

Benade, A. H., *Fundamentals of Musical Acoustics* (New York, 1976)

Blumenfeld, H., *The Syntagma Musicum of Michael Praetorius* (New York, 1980)

Boughley, A. S., *Man and the Environment* (London, 1975) Chaps 2–5

Boult, A., *A Handbook on the Technique of Conducting* (London, 1968)

Buchner, A., *Musical Instruments through the Ages* (London, 1956)

Clark, J. G. D., *World Prehistory* (Cambridge, 1969) Chaps 2–4

Cole, M., *The Pianoforte in the Classical Era* (Oxford, 1998)

Cooke, P., *The Fiddle Tradition of the Shetland Isles* (Cambridge, 1986)

del Mar, N., *Anatomy of the Orchestra* (London, 1983)

de Reuck, A. V. S., and J. Knight, *Hearing Mechanisms in Vertebrates* (London, 1968)

Duifuis, H. J. W., P. Horst, P. van Dijk and S. M. van Netten *Biophysics of Hair Cell Sensory Systems* (Singapore, 1993)

Ellis, A. J., (1885) in Dover edition of Helmholtz (1870) pp. 483ff, 515ff

Ellis, A. J., in *Grove's Dictionary of Music* (London, 1927) Vol. V, pp. 404ff

Hall, D. E., *An Introduction to Musical Acoustics* (Belmont CA, 1980)

The Harvard Dictionary of Music (London, 1946) pp. 752–3

Helmholtz, H., *On the Sensations of Tone* (1870, trans. A. J. Ellis 1885) (Dover edition, New York, 1964)

Houtsma, A. J. M., 'Pitch Perception', *Hearing*, ed. Moore (London, 1995)

Kiang, N. Y. S., *Discharge Patterns of Single Auditory Fibers*. MIT Research Monograph 35 (Cambridge MA, 1969)

Klop, G. C., *Harpsichord Tuning* (Garderen, the Netherlands, 1974)

Krumhansl, C. L., *Cognative Foundations of Musical Pitch* (Oxford, 1990)

Lorente de No, R., 'Anatomy of the Eighth Nerve', *Laryngoscope* 43 (1933) 327–50

Mcgillivray, J. A., 'The Woodwind', *Musical Instruments through the Ages*, ed. A Baines (Harmondsworth, 1961) pp. 237ff

Moore, B. C. J., *Psychology of Hearing* (London, 1982)

Moore, B. C. J., (ed.) *Hearing*, ed. Moore (London, 1995)

Moore, B. C. J., 'Frequency Analysis and Masking', *Hearing* (London, 1995)

The New Oxford Companion to Music, ed. D. Arnold (Oxford, 1983)

Page, C., *Voices and Instruments of the Middle Ages* (London, 1987)

Palmer, A. R., and E. F. Evans, *Hearing*, ed. Moore (London, 1995) p. 77

Pickering, N. C., *The Bowed String* (New York, 1989)

Read, G., *Music Notation* (Boston. 1969) pp. 409ff

Rose, J., J. Hind, D. Anderson and J. Brugge, 'Response of Auditory Fibers in the Squirrel Monkey', *J. Neurophysiol.* 30 (1967), 769–93

Scholes, P., *The Oxford Companion to Music*, 9th Edition (Oxford, 1967) p. 133

Stroux, C., 'The Hydraulos', *Symposium on Music in Ancient Greece* (Athens, 1999)

Wellin, N. L., *Biomusicology* (Stuyvesant NY, 1991)

Werkmeister, A., *Musicalische Temperatur* (1691). Edn. of R. Resch (Utrecht, 1983)

Wever, E. G., *Theory of Hearing* (London, 1949) (Dover Edition, New York, 1970)

Whitfield, I. C., *The Auditory Pathway* (London, 1967)

Wood, A., *The Physics of Music* (London, 1962)

Yasser, J., *A Theory of Evolving Tonality* (New York, 1975)

Yates, G. K., 'Cochlear Structure and Function', *Hearing* ed. Moore (London 1995)

Zhang, J., 'Chinese Bone Flutes', *Nature* 401 (1999), 366–7

Index